Emerging Dynamics: Science, Energy, Society and Values

Loucas G. Christophorou

Emerging Dynamics: Science, Energy, Society and Values

Springer

Loucas G. Christophorou
Academy of Athens
Athens, Greece

ISBN 978-3-319-90712-3   ISBN 978-3-319-90713-0 (eBook)
https://doi.org/10.1007/978-3-319-90713-0

Library of Congress Control Number: 2018942504

Printed on acid-free paper

This Springer imprint is published by the registered company Springer International Publishing AG part of Springer Nature.
The registered company address is: Gewerbestrasse 11, 6330 Cham, Switzerland

To my grandchildren
Isabel, Nicholas,
William, and Loucas

Foreword

The world has been changing always. Leaving aside timescales for the Universe and the geological Earth, we are nowadays witnessing fast changes in our everyday life. In addition, we understand that the world is complex. The basis in understanding complexity is to recognize that everything around us is made of interconnected and interdependent elements (particles, pieces, individuals) depending on the scale. This understanding concerns physical as well as social systems. It is a real challenge to describe such phenomena in an understandable manner without losing the scientific strictness.

This book by Loucas Christophorou faces this challenge. It is written by an excellent physicist who has wide-ranging experience from working in leading laboratories of various countries, but even more than physics has been analyzed. Namely, during his career the author has developed an attentive eye not only in physics but also in looking around in the society. Whether we like it or not, in contemporary society facts do not speak for themselves but communication is needed between scientists and society. This means that scientists have not only to explain facts but to also interpret them for the public sector, especially for policy-makers.

The story in this book starts with a short explanation about the beginning of the universe and then turns to *Homo sapiens* and modern civilization. Then the importance of scientific concepts and the continuous research which explain our world are discussed using many examples. It is clear that the world is an open system where the environment and possible interactions between its members are decisive for the future. The role of scientists in contemporary society is underlined much in line with the saying by Josef Rotblad, the Nobelist: scientists know more and that is why their responsibility in the society is higher. The next logical question about scientific and technological frontiers is then asked and answered by analyzing brain research, molecular biology, and new carriers of energy. The boundaries of science is a fascinating problem, and, although briefly discussed, the problem of possible boundaries involving the essence of extrapolation of scientific knowledge is analyzed.

There are two focal issues in the book which actually stem from the experience of the author. These issues are energy and values. All of our universe and life on the Earth has been and is developing using energy in its various forms. In society, however, values dictate our behavior. The undersigned joins the author stressing the importance of these two notions. Indeed, in physical systems either directly or indirectly, the thermodynamical constraints are of importance while in social systems much is constrained by values. Chapters 5 (on values) and 7 (on energy) are in my view the central ones in the book.

The final chapter is on the future. What will happen, the author asks. It is no surprise that the author stresses the need to be guided by human values, whatever the scientific applications are used in the future. In this context, the principle of complementarity with many components – science, philosophy, art, spiritual world – is strongly advocated. The future is based on science and values, and the undersigned cannot agree with the author more. What makes the explanations valuable is the comparison of Hellenic, Christian, and Western perspectives with the scientific perspective. And the challenge stated at the very end of the book is really optimistic: "A science guided in its applications by human values and a value system cognizant of the facts of science and willing to accommodate them." This is something to be recognized in societal affairs and for evidence-based politics worldwide.

An appendix on energy is a valuable supplement which summarizes the experience of the author in this field. In it he succeeded to explain in a nutshell not only the scientific but also philosophical and theological aspects of energy. What makes the description even more informative are the explanations of terminology based on the mother tongue of the author – Greek.

Loucas Christophorou has written an excellent book for all those who care about the world. He should be congratulated on such a significant advance of intellectual thoughts. The book is especially recommended for reading and understanding to policy-makers locally and globally.

Tallinn, Estonia Jüri Engelbrecht

Preface

This book is about science, science-based technology, and the impact of both on humanity and its future. It is also about science and values and the significance of both for the future of humanity. It is argued that both science and values are prerequisites for a hopeful future.

Modern civilization's most distinct characteristics are due to science, science-based technology, and energy. The role of energy for the sustainability of civilization and the impact of biomedical science on man are especially emphasized throughout the book.

Four themes run concurrently through the entire book: (i) The overwhelming impact of modern science and science-based technology on virtually every aspect of life and on man himself; (ii) human values and their significance for science and society; (iii) the need for mutual accommodation between the values of science and the traditional values of society, based on an open mind and a value system merging toward the common and the complementary; and (iv) the fundamental role of energy for civilization and society.

The book deals with the subject comprehensively; it cuts across scientific disciplines and looks at society and modern civilization through the knowledge provided by science and other means of knowing. Although the term science is used generically embracing all sciences, it largely refers to the physical, chemical, biomedical, and the other branches of natural science. Similarly, the term knowledge is used generically and is treated holistically, encompassing knowledge acquired by deduction, reduction-induction, the experimental method of science, the facts of experience and history, and knowledge obtained by means other than the method and understanding provided by science such as philosophy, the arts and man's faiths and cultural traditions. The term value is also used generically embracing the values of cultures, faiths, and science. The book is unique in its approach.

In this timely book, a case is being made for a hopeful future based on science and values – a science guided in its applications by human values and a value system cognizant of the facts of science and willing to accommodate them.

Athens, Greece

Loucas G. Christophorou

Acknowledgments

I am grateful to many friends and colleagues for discussions on various aspects of this book, and to those colleagues who granted their permission to reproduce figures from their published work. I am also deeply thankful to Professor *Jüri Engelbrecht* who wrote the book's foreword, and to Professor Ivo Šlaus, Professor *Stamatios Krimigis* and Professor *Christos Zerefos* who read the book prior to publication and commented on it. I thank Mrs. Katerina Panagiotakopoulou for her help with the manuscript. To my wife *Erato* and daughters *Penelope* and *Yianna*, I owe a debt of gratitude for their support.

Contents

Chapter 1
Arrows of Time

1.1 From the Beginning of the Universe to *Homo sapiens*

Much has been said, conjectured, speculated and dreamed about the origin and evolution of the universe, by scientists and non-scientists alike. Any system of thought claiming to provide an understanding of the physical world, made some statement about the origin and evolution of the universe.

Today, most scientists accept the theory that the universe did not always exist but came into being some time ago with "a burst of a dense dot of pure energy", a "cosmic explosion", a "big bang", 13.8 billion years ago.[1,2,3,4,5,6,7,8,9,10] In the beginning, all was light and there was nothing material before it. Other less accepted views maintain that the theory of cosmic explosion does not prove that there was a beginning of time, as the present expansion of the universe may be one phase of an oscillating or a cyclic universe.[11] There is also the extreme hypothesis of the "multiverse" according to which our universe could be one of an infinite set (see endnote 10).[12,13]

Modern science affirms that the universe began in featureless simplicity and grows more elaborate with time ever since; it is a universe that makes itself *perpetually*. The physical world we live in today is the present stage of the universe's 13.8 billion-year-old evolution.

The principal scientific evidence that the universe is not eternal, but that it began to exist in a cosmic explosion 13.8 billion years ago is the following: (i) The expansion of the universe, (ii) the existence of cosmic background radiation, (iii) the fact that the universe today is not in thermodynamic equilibrium, and (iv) the relative abundance of the different elements, for instance, hydrogen (H) and helium (He), in the universe today.

(i) Nearly a century ago (in 1929), science discovered that the universe is expanding.[14] Measurements by Edwin Hubble showed that the distant galaxies recede from each other with a speed approximately proportional to the distance between them. Thus, the universe is expanding in all directions and it was denser in the past. Given the expansion of the universe, science arrived at the beginning of the

L. G. Christophorou, *Emerging Dynamics: Science, Energy, Society and Values*,
https://doi.org/10.1007/978-3-319-90713-0_1

universe, at the big bang, starting from today's scientific facts. Based on the physical laws as we know them today, science has arrived, gradually progressing backwards in time, to moments when the universe was denser and hotter, until the moment when the universe was *unimaginably small*, *unimaginably dense*, and *unimaginably hot* (according to some theories, the temperature of the universe in its first 10^{-43} seconds exceeded 10^{32} degrees Kelvin). This moment, 13.8 billion years ago, marks the beginning of the universe. There is, therefore, clear scientific evidence that the universe has a beginning.

Theory contends that in the very first moments after the big bang, elementary particles would have been present in copious numbers in thermal equilibrium, "in a state of continual mutual interaction." Beyond this extremely young universe, the nature of the energy and matter in the universe changed depending on the universe's age and its declining temperature. The big-bang theory offers a consistent account of the history of matter (see endnote 13).

(ii) A little later (in 1965) science discovered that there is *cosmic background radiation* evenly distributed throughout the universe, which today corresponds to a temperature of 2.7 K. The uniform distribution of cosmic background radiation shows that it concerns the entire universe and that it is the radiation which was left over when the universe was still very hot (~3000 K) and very dense and its main constituent was the thermal background radiation. As the universe expanded, the cosmic background radiation corresponded to lower temperatures up to its present value of 2.7 K. The existence of the cosmic background radiation is a clear indication that the universe began to exist at some time in the past. The observations of Arno Penzias and Robert Wilson on the existence of cosmic background radiation uniformly distributed in the universe were announced in 1964. In 1989 NASA launched the "*Cosmic Background Explorer (COBE)*" satellite, which found that the spectrum of the cosmic background radiation coincides almost entirely with that of an ideal black body at a temperature of 2.725 ± 0.002 K. This observation is amazingly consistent with the predictions of the big-bang theory.[15]

(iii) Since the universe contains whatever exists, it constitutes a closed thermodynamic system which tends toward thermodynamic equilibrium. If the universe were eternal, it would already have been degraded energetically and it would have already ceased to exist. Since the universe today exists and it is not in thermodynamic equilibrium, it cannot be eternal, but it ought to have begun to exist.

To the conclusion that the universe is of finite age, one is led also by considering that the energy of radioactive atoms (radioactive nuclei) decreases over time because radioactive atoms are metastable and they decay (are de-excited) automatically, radiating a portion of their energy. If the universe were eternal, there would be no radioactive atoms on Earth today; they would already have been de-excited and they would already have been converted into stable atoms. Similarly, one might observe that if the universe were eternal, the interior (the core) of the Earth would not be hot today; it would already have been cooled down.

(iv) The relative abundance of various atoms: hydrogen, *H* (10,000); helium, *He* (1000); oxygen, *O* (6); carbon, *C* (1); all the rest types of atoms (<1) (see endnotes 4, 9, 10).[16] Hydrogen and helium are primitive elements, they were created mostly in the early phases of the universe and they reveal the characteristics of its evolution. The ratio of *H* to *He* in the universe was largely fixed within the first few minutes of the beginning of the universe and is consistent with observations today (see endnotes 9, 10).

Modern science therefore considers that the cosmic explosion marks the absolute beginning of the physical universe: the absolute beginning of *time, space, energy (matter) and change*. Time started when space started and energy was created; from that moment onward commenced the unceasing perennial change and evolution of the physical universe. The expansion of the universe and the consequent drop in its temperature and density, determined its material composition under the perpetual influence of the forces of nature and the incessant transformations of energy.

In the absolute beginning of the universe the prevailing conditions were extreme. Although we do not know well the forms of energy in the first moments of the universe, we know that in the beginning all was energy, incomprehensible quantity of energy in the form of pure radiation (light)[17] under extremely high temperatures (see endnotes 4, 6–10, 13, 16), and extremely high particle energies.[18]

The radiant energy (light) at the beginning of the universe has gradually been transformed into other forms of energy, other types of radiation, and other types of particles and antiparticles (see endnotes 4, 6–10, 13, 16); at first to *quarks;* a little later to *nucleons* (protons and neutrons) and to *leptons* (electrons, neutrinos and light particles); and much later, to *atomic nuclei* (from the fusion of protons and neutrons). In just the first few minutes of the universe's life, all the essential basic ingredients for creating *neutral atoms* of matter emerged from the primordial radiant energy. Although the atoms of hydrogen and helium appeared in the first few minutes of the universe's age, the *atomic era* followed much later (~50,000 years after the big bang).

When the universe was ~300,000 years old and its temperature ~3000 K, the universe began to fill with neutral matter (see endnotes 4, 9); the electrons and the nuclei that existed began to combine to produce neutral atoms. With the disappearance of the electrons and the nuclei, matter began to become transparent to radiation and light began, ever since, to fill the universe.

Progressively, the energy composition of the universe began to change dramatically; the density of matter began to overtake the density of radiation and ultimately matter (the condensed form of energy) prevailed in the universe. The simple neutral matter (initially in the form of *H* and *He*) became successively more complex and diversified the microscopic and macroscopic composition of the universe. The ceaseless transformations of energy and the resultant perpetual change led to the macroscopic universe, to its wondrous structures, and, on Earth, to the amazing order and organization of biological organisms and to life itself. Ninety-two distinct kinds of atoms occur naturally on Earth and everything on Earth and the universe is made up of these atoms and their combinations. This universe allowed complexity and consciousness to develop on Earth; it was "tune to life" from its inception.

Today, the radiation in the universe (mainly as cosmic background radiation) is a very small percentage of the matter-energy that exists, and antimatter no longer exists on the macroscopic scale (our entire galaxy consists of only matter and not antimatter) (see endnote 13). The matter and antimatter that existed in the initial stages of the universe were by-and-large mutually neutralized under conditions that led to the dominance of matter as we see today. How did matter prevail over antimatter? We do not yet have a complete explanation of this asymmetry between matter and antimatter.[19] Across the universe, about 99.9% of ordinary matter exists in the form of hydrogen (*H*) and helium (*He*) (see endnote 9) while, according to recent discoveries in astronomy and cosmology, *dark matter* seems to prevail over ordinary matter (see endnotes 10, 13).[20,21] The existence and the properties of dark matter are inferred indirectly from the effects of its gravitational field. But what is dark energy? Scientifically we still do not know.[22]

And so, science has led us to a uniquely singular moment, to the absolute beginning of the creation of the universe. Science, however, is not able to explain what caused this beginning and from where the primordial energy in the beginning came.

While much remains uncertain and unknown, the following is clear: *the current state of the universe, and any previous state of the universe since the beginning of time, was brought about by energy and its endless transformations and incessant flow; the energy that was in the beginning of time is today in enormously different forms, distributions, complex structures, and organisms, and its behaviour is explained by a multitude of physical laws.*

From the *initial (primordial, αρχέγονη) energy* at the beginning of the universe came all subsequent forms of energy and matter that have since existed and presently exist. Every new form of energy derives from some other form (or forms) of energy that existed before. Energy comes from energy. It could in fact be said that we live in a physical universe of energy, where everything is a manifestation of the different forms of energy. In this universe, the unceasing transformations of energy degrade the universe's energy and increase its entropy and disorder, while concomitantly they lead to order and organization; everywhere and always, unceasingly, they differentiate the constitution of the universe and account for the physical phenomena and the universe's behaviour and evolution. Energy today is of very different forms and very differently distributed in the universe than it was in the distant past.

In the initial energy that was created at the beginning of time "from nothing" was contained all that was necessary for the evolution of the universe and life. We exist as living organisms; consequently, the universe had all the necessary forms of energy for life, at least on this planet. While energy is the essence of the physical universe, it is difficult to answer adequately the question "*What is energy?*" In a recent book (see endnote 16) instead of answering the question as such, I attempted to bracket it by answering another question, namely, *"What do we know about energy scientifically?"* We know that energy and mass are equivalent, that there are many forms of energy, that energy can be transferred from one state of a system to another, that it can be distributed among the states of a system or among the particles constituting a system, and that energy is degraded from higher to lower values; we also know that energy is transformed from one form to another, that energy facilitates change, perennial ceaseless change; we know, furthermore, that energy is conserved and that it can be stored microscopically and macroscopically, that natural processes follow the flow of energy and that energy is the agent of order and formation of new structures; we know that energy is essential for life and vital for civilization.

The incessant flow of energy caused *all ordering and organization in the universe.* The natural trend toward energy degradation and disintegration can be "locally" reversed by energy input so that new order is generated and new structures are formed and life itself can become possible and sustainable. Through energy input and energy dissipation, energy change and energy transformation, the organization of matter is constantly growing more complex. Energy, thus, becomes the basis of nature's activity and the source of the perennial becoming of the cosmos. There was energy before life appeared on Earth that "prepared" the terrestrial environment for life, and there has been energy in all its various forms to make life possible, including us.

1.2 Increasing Complexity and Disorder

According to the second law of thermodynamics, an increase in *entropy* accompanies every spontaneous natural process. The entropy of an isolated thermodynamic system never decreases; it either stays constant (reversible processes) or it increases (irreversible processes). Since all natural processes are irreversible, the second law of thermodynamics requires that the total entropy, S, of any system plus that of its environment increases ($\Delta S > 0$) as a result of any natural process. Entropy thus always increases with time; it tells us the direction time and natural processes are going. The concept of entropy and the second law of thermodynamics explain why the increase in biological complexity is time asymmetric. *Entropy defines an evolutionary arrow of time from past to the future.*

Ludwig Boltzmann described entropy as a measure of molecular disorder and concluded that the law of entropy increase is simply a law of increasing disorganization – inexorable rise in the level of complexity. Boltzmann described the microscopic meaning of entropy by showing that the entropy S of a system in a given macro state can be expressed as $S = k\ lnW$, where k is the Boltzmann constant and lnW is the natural logarithm of the number of microstates, W, corresponding to the given macro state. The increase in entropy which accompanies every spontaneous natural process, then, means that natural processes tend to move toward a state of greater disorder.

The second law of thermodynamics does not forbid a *decrease* in entropy. The entropy of one part of the universe may decrease in any process because the entropy of some other part of the universe increases by a greater amount so that the total entropy always increases. The decrease in entropy requires energy input to the system and leads to order. Hence, in a world of change and decay there is also order and organization.

Order in Nature, writes Davis,[23] is apparent in two different forms: (i) in spatial patterns of atoms such as in crystals, and (ii) in living organisms. In the former case order is manifested because of the system's simplicity, in the way the atomic arrangement repeats itself in a regular pattern throughout the material, while in the latter, order is manifested because of the system's complexity and "the way its diverse component parts cooperate to perform a coherent unitary function" (see endnote 23). In both cases the concept of order refers to the system as a whole.

Order leads to new structures which are formed and maintained through the supply of energy to the system allowing it to reverse its natural tendency toward increased entropy. Complexity and organization are distinguishing features of living systems that require energy input. Living systems are open to their surroundings and they, thus, "communicate" with their environment; a basic element of this communication is energy exchange[24] and energy flow.[25,26]

Living matter, many have argued, has an innate tendency to self-organize and to auto-replicate, and replication requires a very high threshold of complexity.[27] The study of the complex structures of living organisms bares little similarity to the study of matter by reductionist physics. Complexity is not the result of the combination of many simple processes that occur on a more fundamental level; there are always unknowns beyond the sum of the knowledge provided by reductionist science.[28] Are, then, the complexities of biology reducible to physics or is biological complexity, self-organization, auto-replication, information, etc. not possible to treat that way? If reductionist science is not capable to provide an understanding of such properties of living systems, are there laws at another level of knowledge which are fundamental in the sense that they cannot logically be reduced to the laws of physics? A complex system may possess collective properties that are absent, or meaningless, for its constituent parts; its behaviour may not be possible to reconstruct from basic physics.[29] From another perspective, as Davis writes,[30] it remains a basic scientific *challenge to demonstrate how localized interactions can exercise global control and if this can be explained in mechanistic terms at the molecular level*, that is, to account for biological forms in terms of reductionist physics. The human brain is by far the most complex physical object known to us in the entire cosmos (see Chap. 4).

1.3 Everything Changes and Evolves

1.3.1 Everything Changes

Change began when time and space began. The beginning of change, therefore, coincides with the beginning of the universe. The reason for the perennial change of the physical world is the forces of Nature and the incessant transformations of energy. Change is an intrinsic property of all matter, organic and inorganic, and of all living organisms. Nature changes in every quantum event! The basic units of organic and inorganic matter and of living organisms, although they remain well characterized they perpetually change under the forces of Nature and their interactions with their environment. Nothing remains still. We ourselves are but change incessantly; the same holds for our societies. The perennial change in Nature changes everything, and because everything changes, everything evolves.

1.3.2 Everything Evolves

The term "evolution" usually refers to biological evolution. However, the processes by which planets, stars, galaxies, the universe, life, and societies change over time are also types of "evolution". A 1999 report by the US National Academy of Sciences[31] on biological evolution, sites evidence of evolution from three sources: (i) the origins of the universe, earth and life, (ii) biological evolution, and (iii) human evolution.

1.3.2.1 The Universe Has Evolved

We have already referred to the origin of the universe earlier in this Chapter. It suffices here to say that the universe is engaged in a unidirectional change where the preponderance of the physical processes that occur are irreversible. Independent scientific methods give an age for the solar system of about 5 billion years and an age for our galaxy two to three times greater.[32]

1.3.2.2 The Earth Has Evolved

The age of the Earth is estimated to be 4.54 billion years.[33] The Earth had been shaped by energy over billions of years before it became fit for life.[34] From the atoms of the elements *H* and *He* that were formed in the initial stages of the universe, were built, step-by-step, the rest of the elements making up the cosmos, and from the combinations of those elements, in time, were forged the complex structures of biology.

Of all the chemical elements, the most fundamental for life as we know it on Earth is that of ***carbon***. It has been correctly said that carbon "structures and fuels all of life.... every living thing stores its genetic information in the same language, a chemical alphabet written in a carbon script;" carbon "holds, frees, and remakes the molecules of life".[35] The explanation of how carbon was initially formed in the universe (see endnotes 2, 6, 10, 16)[36] is one of the most fascinating examples of the intricate role of energy in the cosmos and an illustration of how energy features of microscopic matter critically determine macroscopic characteristics of the cosmos, including life itself.

1.3.2.3 Life on Earth Has Evolved

We do not know the origin of life scientifically. However it happened, it must have been the culminating event at the end of a long and complicated process of preparation. The question "*How did life begin?*" can be asked in several other ways; for instance, "*Where did the first organism come from?*", "*How did life arise from dead matter?*", "*Which path of chemical evolution succeeded in initiating life on Earth?*", "*How did the first self-replicating, information-carrying organism arise?*", "*How complex replicating systems have arisen spontaneously?*". The answer to all these questions is: ***we do not presently know.***

While we do not know the origin of life, we do know that life on Earth is exceedingly complex and that since its beginning it is evolving, becoming more complex as it evolves. Life as we know it requires complex chemistry and most of the elements that are used in the chemistry of life were created in the deep interiors of the stars. For these elements to become available for life, they had to be released by the stars in which they were formed. This happened at the end of the lifetime of those stars, which is typically billions of years. Thus, it can be argued that the origin of complex life forms on Earth could not have happened in less than 5–10 billion years after the big bang, since the first generation of stars would not have contained elements like C and O that are necessary for life as we know it.

There is evidence that bacteria-like organisms lived on Earth 3.5 billion years ago, and they might have existed even earlier.[37] Evidence for the existence of more complex organisms (that is, eukaryotic cells which are more complex than bacteria), has been found in fossils sealed in rocks approximately 2 billion years old. Table 1.1 lists the order in which increasingly more complex forms of life appeared (see endnote 31).

The fossil record is seen to provide evidence of systematic change through time – of descent with modification. It can also be seen from the data in Table 1.1 that it took

Table 1.1 Order of increasingly more complex forms of life (see endnote 31)

Life form	Millions of years since first known appearance (approximate)
Microbial (prokaryotic cells)	3500
Complex (eukaryotic cells)	2000
First multicellular animals	670
Shell-bearing animals	540
Vertebrates (simple fishes)	490
Amphibians	350
Reptiles	310
Mammals	200
Nonhuman primates	60
Earliest apes	25
Australopithecine ancestors of humans	5
Modern humans	**~0.15 (150,000 years).**

billions of years to produce multi-cell organisms and several hundreds of millions of years to produce animals with complex brains. A great deal of time elapsed for life to reach sentience and intelligence. The first specimens recognized as modern *Homo sapiens* date from about 195,000 years ago (see endnote 37).

Living organisms have extraordinary information processing capabilities. Today, the accepted view is that all living organisms store and transmit hereditary information using two kinds of molecules: *DNA* (deoxyribonucleic acid) and *RNA* (ribonucleic acid). Each of these macromolecular structures is composed of four kinds of subunits (*DNA*: adenine, guanine, thymine and cytosine; *RNA*: adenine, guanine, thymine and uracil) known as nucleotides. The sequences of nucleotides in lengths of *DNA* or *RNA* are known as genes. The sequence of nucleotides in *DNA* and *RNA* determines the sequence of amino acids in proteins. Every living cell has two dominant kinds of large molecules: proteins and nucleic acids. Proteins act as catalysts to make other chemicals react in highly specific ways, while nucleic acids function "to create and encode and then store information in the nucleus of every living cell of every organism" and, in turn, "to transmit and express that information inside and outside the cell nucleus".[38]

The code to translate nucleotide sequences into amino acid sequences is essentially the same in all organisms. Moreover, proteins in all organisms are invariably composed of the same set of 20 amino acids. This unity of composition and function is an argument in favour of the common descent of the most diverse organisms.

Genetic variations result from changes (mutations) in the nucleotide sequence of *DNA*. Such changes in *DNA* can be detected and described with precision. Slight changes in genes can have big effects. Genetic mutations arise by chance.

As the ability to sequence the nucleotides making up *DNA* has improved, it has become possible to use genes to reconstruct the evolutionary history of organisms. Because of mutations, the sequence of nucleotides in a gene gradually changes over time. The more closely related two organisms are, the less different is their *DNA*. Because there are many thousands of genes in humans and other organisms, *DNA* contains enormous amounts of information about the evolutionary history of each organism.

1.3.2.4 Humanity Has Evolved

The origin of *Homo sapiens* poses questions related to those of the origin of life. The paths that lead from the origins of "*primitive life*" which existed on Earth at least 3.5 billion years ago, to the diversity of life that exists today, are numberless, long, tortuous and still not altogether known and understood.

On the biological level, there are fairly definite ideas about the "genealogy" of *Homo sapiens*. Distinctive bones of the oldest species of the human genus, *Homo*, date back to rock strata about 2.4 million years old.[39] The fossil record shows that the **human** genus first spread from its place of origin in Africa to Europe and Asia a little less than two million years ago (see endnote 39). The first specimens recognized as modern *Homo sapiens* date from ~195,000 years ago (see endnotes 37, 39).[40] It is believed that humans arose from ancestral primates; they did not evolve from modern apes, but humans and modern apes are believed to share a common ancestor, a species that no longer exists. Molecular biology has provided evidence for the close relationship between humans and apes (see endnote 37).

Man, it is argued, can neither be explained by evolution, nor is his conscious self and intelligence in the evolutionary process. The human brain is by far the most complex physical object known. Alfred Russell Wallace maintained[41] that man's brain cannot be explained by natural selection. Even today, scientists (see endnote 23) think that human intelligence refutes Darwinism, and evolutionist Ernst Mayr argues (see endnote 40) that intelligence is not an outcome of evolution. He goes on saying that of the 50 million species on Earth only man created civilization; that human beings with their brain capacity, their use of complex language, and their ability with abstract reasoning, represent the pinnacle of life on Earth, far outdistancing any rivals. Millions of centuries of evolution along numberless paths, have led to this result only in human beings! ***We exist as unique self-conscious intelligent beings outside the evolutionary process and beyond the reductionist view of man***.

However, we are still not able to answer either of the fundamental questions: "*What makes intelligence possible?*" and "*What makes consciousness possible*?" According to Leakey (see endnote 39) the origin of human consciousness has begun within the last 2.5 million years. Leakey also points out that "deliberate burial of the dead", a human activity "redolent of consciousness", occurred about 100,000 years ago. Earlier than that, there is no evidence of any kind of ritual that might signal reflective consciousness, nor is there any evidence of art (see endnote 39). Leakey argues that there is evidence of modern humans ("humans who spoke like us and experienced the self as we do") from about 35,000 years ago (see endnote 39).

1.3.2.5 The Closeness of Life

Science, foremost molecular biology and genetics and especially genomics, expanded profoundly our understanding of life and its history and evolution. It has brought to light a most distinct characteristic of all life: *its closeness.*

The human genome is about 3.1 billion base pairs in length (see endnote 37). There are about 20,000–25,000 protein-coding genes in the human genome and the total amount of *DNA* used by those genes to code for protein is 1.5% of the total. For simpler organisms (e.g., warms, flies) the number of genes is about the same, around 20,000. Clearly, then, gene count must not be the whole story. By any estimation, writes Collins (see endnote 37), "The biological complexity of human beings considerably exceeds that of a ground worm even though the gene count is similar in both. Our complexity must arise not from the number of separate instruction packets, but from the way they are utilized." The code used to translate nucleotide sequences into amino acid sequences is essentially the same in all organisms. Moreover, proteins in all organisms are invariably composed of the same set of 20 amino acids; the proteins of bacteria and the proteins of humans are the same: *impressive degree of molecular structural uniformity*.

To what extent, then, human genes are unique and to what extent are they shared by other organisms? The best estimate is that humans and chimpanzees are 96.6% identical at the *DNA* level and that humans share about half of their genetic instructions with bananas (see endnote 37).[42] This implies that while there are vast volumes of genetic information that keep human cells running (and these are shared throughout the living world), it is the tiny differences within commonly used genes, plus the small percentage of unique genes, that give human beings their uniqueness (see endnotes 37, 40).

Minor alterations in the fine detail within the genome give individuality within the species. A comparison of the genome of different humans has shown that at the *DNA* level, humans are 99.9% identical and that similarity applies regardless of which two individuals from around the world are compared (see endnote 37). A tiny fraction (0.1%) of the human *DNA* differs from person to person. This remarkably low genetic diversity distinguishes humans from other species whose amount of *DNA* diversity is 10 or even 50 times greater (see endnote 37). It should, however, be observed that several biological studies show[43,44,45] that genes are only one aspect of our inherited make-up and that any assessment of how those genes are put to work to create each individual person needs careful consideration.

Genes become altered or "mutated", albeit very slowly. The copying of *DNA* is, as a rule, extremely faithful. In people and other multi cellular organisms, on average, a mistake is made only once for every 100×10^6 or so nucleotides of *DNA* copied in a generation. However, because of the large number of nucleotides in a cell's genome, mistakes happen often on a per-cell basis.[46]

Naturally occurring mutations are random; they are of several types (Table 1.2) and generally have small effects. The simplest way to alter a protein is by point mutation, where one amino acid is substituted for another at a position in a protein. How often do random mutations occur? According to Behe (see endnote 46), substitution, deletion and insertion mutations, and gene duplications are estimated to occur at a rate of about one error every 100 million base pairs per generation. Random mutations in biology degrade rather than enhance the complex adaptedness of organisms; most mutations are harmful.

The study of genomes offers new insights into other related subjects of fundamental significance. We shall focus just on two: common descent and evolution.

Table 1.2 Varieties of *DNA* mutations (see endnote 46)

Type of mutation	Description
Substitution	Switch one kind of nucleotide for another
Deletion	Omission of one or more nucleotides
Insertion	Addition of one or more nucleotides
Inversion	"Flipping" of a segment of *DNA* double helix
Gene duplication	Doubling of a region of *DNA* containing a gene
Genome duplication	Doubling of the total *DNA* of an organism

1.3.2.6 Common Descent

Common descent attempts to account for the similarities between organisms. It simply says that certain shared features were there "from the beginning" – the ancestor had them. *DNA* sequencing experiments show that some distinctly related organisms apparently share arbitrary features of their genes that seem to have inherited from a distant common ancestor. If two kinds of organisms share what seems to be a common mutation or set of mutations in their *DNA*, it can be assumed that a common ancestor of the two species originally suffered the mutation and the descendants inherited it (see endnote 37).

The study of genomes leads to the conclusion that humans share a common ancestor with other species. The same mistakes in the same gene in the same positions of both human and chimp *DNA* were found (see endnote 46). If a common ancestor first sustained the mutational mistakes and subsequently gave rise to those modern species, this would account for why both species have them now. The unity of composition and function of the most diverse organisms is a powerful argument in favour of common descent.

1.3.2.7 Evolution

The accumulation of information provided by studies of the genomes of organisms makes possible the contemplation of the mechanism of evolutionary change at the molecular level. Let us look at the implications of this knowledge on the claims of Darwin's theory that random genetic accidents (random mutations) and natural selection working over extended periods of time yield results that do not look like the effects of chance and can, thus, modify life in important ways that account for the differences between organisms.

There are today distinguished scientists who do not doubt the fact of evolution but question the adequacy of the mechanism of random mutation and natural selection. It is argued that to get a realistic understanding of what random mutation and natural selection can do, one must follow changes at the molecular level.

Random mutations Recent knowledge of the sequences of many genomes, of how mutations occur and how often, allows exploration of the possibilities and limits of the random mutation hypothesis. Today, we know that the variations arise at the level of the *DNA* coding and that they are occurring all the time, randomly, in the cells of an organism's body. The claim that the evolutionary change is driven by random mutations is still met with scepticism. "How can chance alone be responsible for the emergence of completely new and successful structures?" is a question often been asked. Similarly, it is pointed out that random mutations tend to degrade rather than to enhance the complex and intricate adaptedness of organisms; most mutations are harmful.

Natural selection There are two prerequisites for natural selection to occur: sufficient variation and enough time. By itself the idea of natural selection simply states that "the more fit organisms will tend to survive". However, as many have noted, the question is not "who will survive", but "how do organisms become more fit". Now that knowledge about the molecular foundations of life is emerging, the question is "Where the complex, coherent molecular machinery did come from?" Michael Behe (see endnote 46) believes that the answer will not involve random mutation at the center. According to Michael Ruse, Stuart Kauffman, Murray Gell-Mann and others, natural selection was rejected "because the scientific evidence failed to convince".[47]

It thus seems that many scientists are not convinced that the complex structures of biology are likely to have resulted from purely random accidents, a mechanism that fails to explain the evolutionary arrow of time.

1.4 An Historic Perspective

1.4.1 Physical, Biological and Structural Uniformity of Living Organisms

Physical science has patiently and systematically unraveled the physical uniformity of the cosmos: everything everywhere is made up of the same type of atoms. In fact, 99% of living organisms are made up of only four (*C*, *H*, *O* and *N*) of the 92 natural chemical elements (atoms).

Science has, also, unraveled the biological uniformity of life: the simplicity and generality of the genetic code. The genetic code by which information in *DNA* and *RNA* is transplanted into proteins is universal in all known organisms. Similarly, science has shown the structural uniformity of living organisms: the stability of their macromolecular biological structures.

The theory of heredity as such deals with the stability of the genes as they are passed from an individual organism to its offspring. Many see evidence in this that all life is a continuum with no precise break between humans and other animals. Others consider that this (physical and biological) uniformity suggests that the difference between humans and other animals must be searched for elsewhere, not in *DNA*. Will modern biology be able to tell us what life is, what is responsible for it or what it is for? Perhaps, but there is a lot we do not presently understand, and perhaps never will without searching beyond science.

Indeed, besides the physical world we live in and are a part of, there is a non-physical world which contains entities that cannot be studied or measured using the instruments of science, and knowledge that cannot be acquired by the methods of science, for they lay beyond science: the knowledge we have through the self and consciousness; our feelings, emotions and perceptions; our love, devotion, friendship, or our awe of the sacred. This largely private world and this kind of largely private knowledge is beyond the boundaries of science.

There is also a cultural world we know which contains all our past and present cultural and intellectual traditions, both beautiful and ugly. "The past is the road by which we have arrived where we are…. No road into the present need be repudiated and no former way of life forgotten", wrote Margaret Mead.[48] Indeed, as she writes, the creeks of primitive cultures, flowing down mountain tops, merged into the rivers of wider civilizations and are now merging into the open seas of a unified humanity headed together for the vast ocean.

In a lecture at the Academy of Athens, I expanded on Mead's theme and wrote[49]: "The uniformity of the ocean lacks the beauty and the bright colors of the creeks and the rivers of its origin, but the ocean still has the color of the sky, the shades of the blue. Will the ocean remember the beauty of its origins and the wondrous journey of the creeks and the rivers that led to it? Will anyone remember where they came from? Or is the question irrelevant? Will the ocean always reflect the color of the sky and will the sky remain blue? I do not know. What I do know is that change will continue and with it the rivers and the ocean."

1.4.2 Societal Complexity

Human society, history tells us, is moving toward higher levels of complexity: toward larger settlements supported by increasingly larger, more complicated and more complex infrastructures; more institutions, more social needs, more specialization; larger information and communications loads, and more societal interconnections through an ever elaborate web of systems and technologies. Increasingly, modern society becomes more organized, more socio-politically controlled, more dependent on powerful technologies to support the services demanded by its population's traditional needs and new habits such as explosive growth in consumer, business and government e-services. The cost of maintaining this societal complexity is becoming more difficult to afford, principally because it requires: (i) processing enormous amounts of energy and information in an increasingly less efficient manner, and (ii) a technological infrastructure which grows ever more complex that becomes difficult to understand and to control. Societal complexity and its maintenance, it is argued,[50,51,52,53] destabilizes society's institutions and diminishes their adaptive capacity. The continuous increase in complexity of modern civilization makes it fragile and operationally vulnerable.

Once complex societies are disrupted, they become unable to support their complexity, they crumble and unavoidably collapse[54]; in the present age of globalization, modern societies might not collapse alone, in isolation. Yet, all indications are that present complex societies will become even more complex in the future. They will require more efficient infrastructure, new technology, information processing and energy supply systems each of which has historically become progressively more complex; for example, every new source of energy (e.g., fossil fuels, nuclear, renewable energy sources) introduced a whole new level of complexity. This complexity extends into the relations of science and society (see Chaps. 3, 4, and 5).

Jared Diamond[55] in a New York Times article titled "The ends of the world as we know them" remarked that "History warns us that when once-powerful societies collapse, they tend to do so quickly and unexpectedly." Complex systems collapse when they have no way to get simpler other than to collapse. If then we want to save civilization – and we do – it would not be enough to stabilize population and energy consumption, but to abandon 'economic growth' and 'progress' defined in terms of complexity or size or power (see endnote 54). Civilization as we know it can become unstable, because too many of its functions are increase-only. In his book "The Collapse of Complex Societies", Joseph Tainter (see endnote 50) argues that societies can reach and pass a point of diminishing marginal returns to investment in societal complexity. A stage can be reached when machines are interacting and trading with each other with little human involvement enabling a more interconnected but less comprehensible technological infrastructure, and a "confusing apersonal high-tech world".[56] Adaptation, it has been argued, can be an antidote to collapse, and adaptability can be an asset for survival. Yet, paradoxically, the greatest threat to the quality of life is that the human species is so immensely adaptable that it can survive under utterly objectionable conditions. Healthy adaptation whether of governments, businesses, social organizations or institutions, needs innovation and almost all innovations can cause both benefit and harm.

Today, powerful new realities challenge ethics in a most fundamental way: man is getting ready to modify and to remake himself and all the rest. We are headed for actions beyond "all former ethics" and we may wonder if we would care about our former ethics, values and the things we were!

Historians usually date the beginning of modern era at the end of the fifteenth century. The modern era has been clearly Western and the Modern Western civilization has been uniquely scientific. Since about the end of the twentieth century, however, many believe that we have entered the post-modern era "with most of the Western traditions abandoned by the elites of Western societies".[57] Clearly, if Modern Western civilization is uniquely scientific, the "post-modern" one will be even more so. Human history claims Zakaria,[58] is not ending, as it has been advocated,[59] it is accelerating, and the reason for this acceleration is science and science-based technology. Profound changes accompany this acceleration that promise hope for a better society and concomitantly instill fear of profound changes in man and society altering both irrevocably. These are discussed in the following Chapters of the book.

References and Notes

1. There is an extensive bibliography in what concerns the cosmic explosion and the initial stages of the evolution of the universe. References 2–10 below are particularly concerned with the form of energy and matter and their evolution during the initial stages of the universe.
2. John D. Barrow and Frank J. Tipler, *The Anthropic Cosmological Principle*, Oxford University Press, Oxford 1986.
3. G. Contopoulos and D. Kotsakis, *Cosmology – The Structure and Evolution of the Universe*, Springer-Verlag, Berlin 1987.
4. Steven Weinberg, *The First Three Minutes*, updated edition, Basic Books, New York 1993.
5. Stephen Hawking and Leonard Mlodinow, *A Briefer History of Time*, Bantam, New York 2008.
6. John D. Barrow, *The Origin of the Universe*, Basic Books, New York 1994.
7. Paul Davies, *The Last Three Minutes*, Basic Books, New York 1994.
8. Brian Greene, *The Elegant Universe*, Vintage Books, New York 2000.
9. Simon Singh, *Big Bang – The Origin of the Universe*, Harper Perennial, New York 2004.
10. Paul Davies, *Cosmic Jackpot – Why Our Universe is Just Right for Life*, Houghton Mifflin Company, New York 2007.
11. See, for example, Ian G. Barbour, in *Physics, Philosophy and Theology*, Robert John Russell, William R. Stoeger, S. J. and George V. Coyne, S. J. (Eds.), Third edition, Vatican Observatory-Vatican City State 1997, pp. 22–48. See, also, William Lane Craig and Quentin Smith, *Theism, Atheism and Big Bang Cosmology*, Clarendon Press, Oxford 1993.
12. Paul J. Steinhardt and Neil Turok, *A Cyclic Model of the Universe*, Science **296**, 24 May 2002, p.1436.
13. Martin Rees, *Before the Beginning – Our Universe and Others*, Basic Books, New York 1997.
14. See, for instance, Reference 10 and John D. Barrow, in *Physics and Our View of the World*, Jan Hilgevoord (Ed.), Cambridge University Press, New York 1994, pp. 38–60.
15. http://lambda.gsfc.nasa.gov/product/cobe/.
16. Loucas G. Christophorou, *Energy and Civilization*, Academy of Athens, Athens 2011.
17. The term "radiation" is often used to include electromagnetic waves of all wavelengths (of all energies). The term "radiation" is used also to include other types of particles besides photons.
18. Cosmic gamma rays with extremely high energies ($>5.7 \times 10^{19}$ eV) have been detected and are believed to have originated in the early universe (Bertram Schwarzschild, Physics Today, May 2010, pp. 15–18).
19. See, for example, Reference 13 (pp. 154–158) and Reference 4.
20. http://en.wikipedia.org/wiki/Dark_matter.

21. There is scientific evidence that ordinary matter constitutes only a few per cent of the mass of the universe and that the majority (96%) of the mass of the universe consists of *dark matter* (23%) and *dark energy* (73%). What is dark energy? Presently we do not know.
22. Some scientists (see, for example, K. Zioutas, D. H. H. Hoffmann, K. Dennerl, and T. Papaevangelou, Science **306**, 26 November 2004, p. 1485) argue that the explanation of dark matter must be sought in "exotic" elementary particles.
23. Paul Davis, *The Cosmic Blueprint*, Templeton Foundation Press, Philadelphia 1988.
24. An organism is constantly exporting entropy by respiration and excretion.
25. Stuart A. Kauffman[26] sees the living cell as a closure of work tasks that propagates its own organization of processes and refers to what he calls "the propagating organization process", which he explains as follows:

    "The release of energy must be constrained into a few degrees of freedom to create work. And it takes work itself to construct those very constraints on the release of energy that then constitutes further work.... It takes work to constrain the release of energy, which, when released, constitutes work.... This is part of what cells do when they propagate organization of process. They have evolved to do work to construct constraints on the release of energy that in turn does further work, including the construction of more constraints on the release of energy. Thus, a self-propagating organization of process arises in cells.... (A living cell) is a closure of work tasks that propagates its own organization of processes."

    This propagating organization of process is not deducible from physics.
26. Stuart A. Kauffman, *Reinventing the Sacred,* Basic Books, New York 2008, pp. 91–94.
27. *DNA* is nature's best-known self-assembling, self-replicating superstructure.
28. Peter Coveney and Roger Higfield, *Frontiers of Complexity*, Ballantine Books, New York 1995.
29. Λουκάς Γ. Χριστοφόρου, *Η Επαγωγική Μέθοδος της Φυσικής Επιστήμης (Από τα Μόρια στον Άνθρωπο;)*, Πρακτικά της Ακαδημίας Αθηνών, τ. 82 Α΄, Αθήνα 2007, σελ. 1–30); Loucas G. Christophorou, *The Inductive Method of Science (From Molecules to Human?),* Proceedings of the Academy of Athens, Vol. 82 A΄, Athens 2007, pp. 1–30.
30. Reference 23, p. 104.
31. National Academy of Sciences, *Science and Creationism – A view from the National Academy of Sciences* (Second edition), National Academy Press, Washington DC 1999, ISBN: 0-309-53224-8.
32. https://en.wikipedia.org/wiki/Solar_System.
33. https://en.wikipedia.org/wiki/Age_of_the_EarthMilky_Way.
34. The process by which the planet Earth prepared itself for life is referred to as the pre-biotic evolution.
35. Eric Roston, *The Carbon Age*, Walker & Company, New York 2008, p. 2.
36. Fred Hoyle, *The Universe: Past and Present Reflections*, Annual Review of Astronomy and Astrophysics **20**, 1982, pp. 1–36.
37. Francis S. Collins, *The Language of God*, Free Press, New York 2006.

38. https://en.wikipedia.org/wiki/Nucleic_acid.
39. Richard Leakey, *The Origin of Humankind*, Basic Books, New York 1994.
40. Ernst Mayr, *What Evolution is,* Basic Books, New York 2001.
41. Alfred Russell Wallace, quoted in Steven Pinker, *How the Mind Works*, W. W. Norton & Company, New York 1997, pp. 299–300.
42. New Scientist, 1 July 2000, pp. 4–5.
43. Reference 37 and Nature-Editorial **518**, 19 February 2015, p. 273.
44. Michael J. Behe, *The Edge of Evolution – The Search for the Limits of Darwinism*, Free Press, New York 2007.
45. H. Pearson, *Genetics: What is a Gene?,* Nature **441**, 25 May 2006, pp. 398–401.
46. Reference 44, pp. 66–73.
47. Michael Ruse, *The Evolution-Creation Struggle,* Harvard University Press, Cambridge, Massachusetts 2005; Michael Ruse, *Is Evolution a Secular Religion?*, Science **299**, 07 March 2003, pp. 1523–1524.
48. Margaret Mead, *Culture and Commitment – A Study of the Generation Gap*, Doubleday & Company, Inc., Published for The American Museum of Natural History, Natural History Press, Garden City, New York 1970.
49. Λουκάς Γ. Χριστοφόρου, *Αέναη και Κρίσιμη Αλλαγή*, Πρακτικά της Ακαδημίας Αθηνών, τ. 84 Α', Αθήνα 2009, σελ. 109–132; Loucas G. Christophorou, *Perennial and Critical Change,* Proceedings of the Academy of Athens, Vol. 84 A', Athens 2009, pp. 109–132.
50. Joseph A. Tainter, *The Collapse of Complex Societies*, Cambridge University Press, Cambridge 1988.
51. M. Mitchell Walbrop, *Complexity*, Simon and Schuster, New York 1992.
52. Robert M. May, *Stability and Complexity in Model Ecosystems*, Princeton University Press, Princeton 2001.
53. Λουκάς Γ. Χριστοφόρου, *Προκλήσεις του Σύγχρονου Πολιτισμού προς την Επιστήμη και τις Αξίες*, Πρακτικά της Ακαδημίας Αθηνών, τ. 89 Α', Αθήνα 2014, σελ. 201–231; Loucas G. Christophorou, *Challenges of Modern Civilization to Science and Values,* Proceedings of the Academy of Athens, Vol. 89 A', Athens 2014, pp. 201–231.
54. Jared Diamond, *Collapse: How Societies Choose to Fail or Succeed*, Penguin Books, New York 2006.
55. Jared Diamond, *The Ends of the World as We Know Them*, The New York Times (1 January 2005).
56. Todd G. Buchholz, *The Price of Prosperity*, HarperCollins Publishers, New York 2016. The author gives examples of poor countries that have fallen apart recently, but also examples of rich countries that have fallen apart as well. See also Mancur Olson, *The Rise and Decline of Nations*, Yale University Press, New Haven 1982.
57. James Kurth, *Western Civilization, Our Tradition*, The Intercollegiate Review-Fall 2003/Spring 2004, pp. 5–13.
58. Fareed Zakaria, *The Post-American World*, W. W. Norton & Company, Inc., New York 2008.
59. Francis Fukuyama, *The End of History and the Last Man*, Free Press, New York 1992.

Chapter 2
Distinct Characteristics of Modern Civilization

Modern civilization's most distinct characteristics are due to science and science-based technology. In this Chapter, we exemplify these characteristics focusing on the prevalent impact of science and science-based technology on man, society, the environment and climate change, and on the fundamental role of energy in both science and society.

2.1 The Prevalent Impact of Science and Science-Based Technology

Science and science-based technology have accelerated the pace of change and innovation and have unified the world; they brought us together – there is no "them" anymore; the boundaries of national civilizations and cultural-value-systems are being blurred. Science and science-based technology enabled the formation of societal infrastructures and modifications to the environment which are vital for the survival and the wellbeing of humanity; they made it possible for more than seven billion people to inhabit the Earth and they improved immeasurably the quality of human life through more goods and services (household appliances, electrification, healthcare, new means of transportation and communication, the ubiquity of the computer and the Internet, to mention just a few) and a myriad of other ways; they helped humanity achieve social justice, freedom and emancipation in many parts of the world and made possible the penetration and the breakup of the "iron curtains" of totalitarian states, liberating oppressed peoples.

Yet, injustice and suffering abide the world over, totalitarian states still enslave their people, and basic human needs for food, energy, shelter, and healthcare are still not satisfied for billions of people, especially in the rural areas of impoverished countries. Humanity is under immense pressure by the billions of the "left-outs". Totalitarianism, terrorism, extremism and war still inflict pain and misery on a grand

L. G. Christophorou, *Emerging Dynamics: Science, Energy, Society and Values*,
https://doi.org/10.1007/978-3-319-90713-0_2

scale the world over, and uncontrolled capitalism and failed government policies lead to unprecedented world-wide economic crises setting humanity back on a slower pace, homogenizing people in their degradation. An unrestrained consumer society lives beyond its means and strains resources and the planet.

Will science-based technology help the poor and will it contribute to social justice? Will means for access to high-technology medicine and communications become broadly available and affordable leading to a better quality of life for all? Will human values and ethics drive science-based technology and push it towards the necessities of the poor by, say, a combination of cheap solar energy, genetic engineering, communications and the Internet? Will humanity be free at last from the terror of the weapons of mass destruction or will it slip deeper into it by *the knowledge of mass destruction*, which magnifies the terror?

Regrettably, powerful new science-based technologies often take off and run away full speed before they are sufficiently understood and adequately assessed for their possible short- and long-range adverse effects on humans and the environment. Time and again, this seems to be the case and it is briefly illustrated in this Chapter for a few new areas such as genetically-modified organisms (GMOs), synthetic biology, and biomedicine. In these and in other fields, the positive and the negative impacts of science-based technology go together and can be profound. The *dual aspects of the impact of science and science-based technology on society and on man himself* will continue and are, in fact, anticipated to intensify in the future; they are exemplified in this Chapter for two areas: ***biomedicine and energy.***

Science-based technological frontiers in biomedicine and energy, and their possible impact on society and on man himself are profound today and they will be more so in the future. The frontiers in biology and medicine will give humanity new powers to treat, prevent and cure disease and to effect beneficial genetic modifications of plants and animals vital for society's future (for instance, increase food production). Simultaneously, these same powers raise new ethical and social issues and fears that emerging scientific and technological frontiers in biomedicine, will determine, in the non-too-distant future, the fate of humanity. Similarly, frontier science-based energy technologies promise abundant, "clean" energy, intelligently conditioned to the needs of modern technology; energy will impact all future functions of society and its availability and affordability will be considered a basic human right. Simultaneously, energy production and use will continue to raise fundamental challenges and serious concerns about its adverse impact on the environment and climate change.

Where, then, are we headed? What are the likely future scientific and technological frontiers and what is anticipated to be their impact – positive and negative – on man, life in general, and climate change and the environment?[1]

Undoubtedly, there will be many new future avenues to knowledge and its use and misuse, and hence enormous shared responsibility by both scientists and non-scientists. This responsibility must be grounded on basic human values and the mutual accommodation of science and society through enhanced dialogue and mutual trust. In our view, the *ultimate future challenge of civilization will be the protection of humanity and the respect of human dignity.*

2.1.1 Biology, Medicine and Biotechnology

In the previous century, we have witnessed the merger of chemistry with physics and gradually the merger of biology with both chemistry and physics. By the end of the twentieth century we have begun to see the gradual reduction of parts of medicine to atoms, molecules and genes, and the beginning of the remarkable explosion in molecular and genomic medicine, driven in part, by *bioinformatics* (the use of computers to rapidly scan and analyze the genomes of organisms). Basic elements of these emerging technologies are the next generation of genome sequencing, genetic engineering, and big-data-driven medicine. In the manipulation of the very small lies new fundamental knowledge for understanding the behavior of the very large, which, undoubtedly, will lead to new technological frontiers in biology, biotechnology, and medicine giving humanity new powers to treat, prevent and cure disease, and to effect beneficial genetic modifications of plants and animals. Concomitantly, these same powers have the potential to change us: the way we are, the way we think about ourselves, and the way we relate to the rest of life and nature. Indeed, in the field of genomics, we do not wish to see the scientists remorseful again, after the fact, repeating what they said after the first atomic bomb explosion, that "the physicists (the scientists) have known sin".

Examples of the new frontiers in these fields are the following:

- *Molecular and genetic roots of cancer*: The processes leading to the development of cancer are extraordinarily complex and there are many diverse types of cancer. If the uncontrolled growth of cells is caused by genetic abnormalities in the cells, then hitting cancer at its molecular origin is of utmost importance. It is generally believed that soon it would be possible to cure many genetic diseases that are caused by the mutation of a single gene. In the case of cancer, one is likely to be dealing *with multigene processes.*[2,3]
- *Stem cell technology*: Stem cells can change into any type of cells in the body and embryonic stem cells retain this ability to re-grow any type of cells throughout their life. Stem cells have the potential to cure diseases such as diabetes, heart disease, Alzheimer's, and Parkinson's. They are, however, controversial and they raise ethical questions because an embryo must be *sacrificed* to extract these cells.
- *Designer genes*: In time, it will be possible to go beyond just the fixing of "broken" genes to enhancing and improving them. Whether designer genes should be used to change the way we look, the way we feel, to make us healthier or something else, we are faced with profound ethical issues.
- *Germline gene modification*: Here one alters the genes of the sex cells and the resultant genes are passed on to the next generation. This is a frontier field, full of promise and peril; replete with scientific, social and ethical concerns.[4]
- *Synthetic biology*: This new field began to surface at the turn of the previous century; it has been described as "the application of science, technology and engineering to facilitate and accelerate the design, manufacture and/or modification of genetic materials in living organisms"[5]; "to create life itself from non-living

materials.... to design living things that meet the specific needs and wishes of humans".[6] According to Cho and Relman,[7] synthetic biology refers "to the creation of synthetic biological systems that are programmable, self-referential, and modular". From its beginning, synthetic biology has been steeped in controversy regarding its potential for societal benefit or harm. Opinions vary from praising synthetic biology for "engineered future life" to how it could lead to the devaluing of life. Unquestionably, the ethical issues raised are monumental.[8]

- *Epigenetics*: Emerging science focusing on changes "in the regulation of gene expression that can be passed on to a cell's progeny but are not due to changes to the nucleotide sequence of the gene"[9]; they are epigenetic (non-genetic) modifications to the genome "that crucially determine which genes are expressed by which cell type and when" (see endnote 9).
- *Human genetics*: The genetic changes that help separate humans from chimps are likely to be profound despite the oft-repeated statistic that only ~ 1.2% of human *DNA* differs from that of chimps. A complete understanding of uniquely human traits will, however, include more than DNA (see endnote 9).[10,11] The *sequencing of the human genome* gives humanity new powers to treat, prevent and cure disease. At the same time, it raises profound new ethical questions and social issues mainly caused by the possibility of crossing boundaries between species. What changes in man? Will, for instance, man proceed and create synthetic forms of life and should he concede rights to non-human animals?
- *Prosthetics*: It will profoundly impact the healthcare and delivery systems. Future robotic prosthetics which mimic what the human body does naturally are being envisioned, and nano-robots might become a reality and might change society profoundly (see endnote 3).
- *Genetic modifications of plants and animals*: Genetically-modified organisms (GMOs) have been applied to plant and animal food resources, and genetically-modified foods (GMFs) are a reality. The benefits – real and potential – of transgenically-modified plants and animals include food supply, enhancement of nutrient security, targeted health such as diet-related chronic diseases, as well as improving herbicide or disease resistance, or drought tolerance. Currently, commercialized GM crops include maize, soya beans, cotton, canola, squash, papaya, sugar beet, tomato and sweet pepper, which are grown primarily in North and South America, and South and East Asia. In efforts to boost agricultural productivity in the world's poor regions, attention has been drawn to Africa.[12,13] Africa, many argue, needs to embrace technologies that enable production of more and better food, and GMOs may increase cereal production especially in Sub-Saharan Africa. However, coexisting with the benefits of genetic modification of plants and animals are known and unknown risks such as possible health risks and food safety, but also possible effects on the environment and socio-economic and ethical issues connected with control of agricultural biotechnologies and intellectual property rights (see endnotes 12, 13).[14] Partly for these reasons, there remains scepticism over GMFs and the issue still divides the EU.[15,16] Despite these concerns, humanity would likely take full benefit of the new age of molecular biology and biotechnology for food production, and would explore further options involving highly polygenic traits.[17]

2.2 The Fundamental Role of Energy

2.2.1 The Significance of Energy for Civilization

In a recent book,[18] I have discussed the fundamental significance of energy for civilization. Energy crucially defines and constrains progress; it sustains civilization. Energy has played a key role in the social and cultural development of humanity. The discovery of new *usable* forms of energy and the development of new energy technologies for the needs of modern civilization introduced important characteristic social changes, which are related directly or indirectly to the production and use of energy.

Energy is prevalent; it has been that way since the beginning of time. Physics tells us that if the big bang is the beginning of space and time, it is also the beginning of energy. From the primordial energy in the beginning, emerged everything that is – the cosmos, the Earth, the environment, us (see Chap. 1).

Today, we know many things about energy, foremost that energy appears in various forms and that it changes, it is transformed, from one form to another – for instance, from electrical energy into light or heat, or from electromagnetic energy into chemical (as in photosynthesis) or electrical (as in photovoltaics). Every energy transformation produces a quantity of low-grade "exhaust" energy normally in the form of heat.

Energy has played a key role in the development of human society and it is the key for a sustainable civilization and a better quality of life. Indeed, the evolution of human society and civilization parallels man's ability to discover and to master new sources of energy. Humanity's need for energy sources that were independent of time and place led to the discovery of new energy sources and none was discarded; humanity used every source of energy it had discovered.

All known forms of energy are necessary for life. We and every other form of life on this planet are inextricably linked to and depend on energy; everything we do relate to energy transformation and energy flow. However, it took millennia before man recognized this fundamental role of energy. The recognition of the importance of energy in science and in society has been slow in coming and it is still unfolding.

Man's reliance on external sources of energy began when he discovered fire and learned to control it and to use it; only then was he able to break the bonds of his physical limitations. So significant was this event that the ancient Greeks created the myth of Prometheus, who stole fire from the gods, carried it to Earth and taught humanity its use (Fig. 2.1).[19] For this act, the myth says, Prometheus was severely punished by Zeus. He was chained to a rock where each day a vulture came to eat his liver away, which was made up again each night, to be eaten away yet again each successive day. An unending torture and a perpetual agony! A severe punishment for a severe crime! A powerful myth for the origin of this "primitive" external source of energy and the significance attached to it by God and Man alike!

Since then, man's need for wood as fuel increased, particularly when he learned to work with metals. Thus, while for thousands of years human labor, the beast, the water, and the wind were used as sources of energy, gradually the use of wood

Fig. 2.1 The picture of Prometheus as torchbearer offering humanity the gift of fire, decorating the Main Hall of the Academy of Athens, Athens, Greece; it is the work of the Austrian artist Christian Griepenkerl (see endnote 19)

became critical and is responsible for the deforestation of large regions of the Earth. In pre-industrial Europe, the forest played the role that would later be assumed by coal and oil. Fossil fuels, were known to man for more than 3000 years (for instance, in China coal was used for heating and natural gas for lighting before 1000 BC).[20,21] However, fossil fuels did not become the main source of energy for humanity until about the middle of the eighteenth century when the discovery of the steam engine signaled the beginning of the industrial base of modern civilization. This is also the beginning of the critical shift made by society toward the science and technology of energy which profoundly changed man's relation to energy and commenced the ascendancy of modern civilization, grounded on science, technology and increased consumption of energy, consisting, up until recently, largely of cheap and abundant fossil fuels.

Surprisingly, while the word "energy" (ενέργεια) was introduced by Aristotle[22] in the fourth century BC and entered the philosophical/theological debates of Christianity in the fourth century AD (see Appendix), the word "energy" did not enter the vocabulary of science and technology until 1807.[23,24] The scientific principles that govern energy were not established until the mid 1800s[25] and even these principles had to be modified when it was discovered that mass was a form of energy. Energy has been usually defined as the equivalent of, or as the capacity for, doing work.

The discovery of the steam engine signaled the gradual transition from wood as a source of energy to fuels of increasingly higher energy density: *wood, coal, oil, natural gas, and uranium* (Table 2.1).

Table 2.1 Energy density of various energy sources (see endnote 18)

Fuel	Energy density (kWh/kg)[a]
Fire wood	4.4[b]
Coal	8.0[b]
Oil	12.0[c]
Natural gas	14.9[b]
Natural uranium (fission of only U_{235})	160,000[c]
Natural uranium (100% fission in a breeder reactor)	22,800,000[c]

[a]1 kWh = 3.6 MJ

[b]David J. C. MacKay, *Sustainable Energy – Without the Hot Air*, UIT Cambridge Ltd., UK 2009, p. 199 (ISBN: 978-0-9544529-3-3)

[c]Γεώργιος Γιαδικιάρογλου, στο βιβλίο Πυρηνική Ενέργεια και Ενεργειακές Ανάγκες της Ελλάδος, Επιμέλεια Λουκά Γ. Χριστοφόρου, Επιτροπή Ενέργειας της Ακαδημίας Αθηνών, Ακαδημία Αθηνών, Αθήνα 2009, σελ. 169. George Yadigaroglu, in Loucas G. Christophorou (Ed.), *Nuclear Energy and Energy Needs of Greece*, Energy Committee of the Academy of Athens, Athens 2009, p. 169

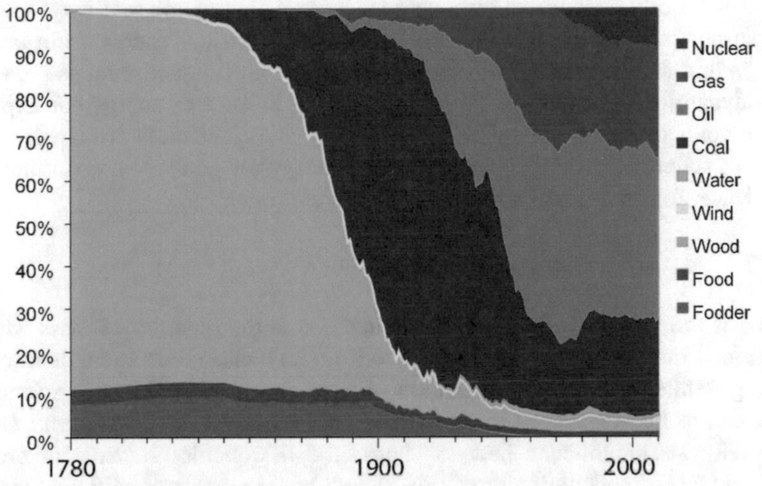

Fig. 2.2 Gradual evolution of the changes in energy sources used by man in the time period from 1780 to 2010, for the USA (see endnote 26)

Figure 2.2 shows[26] this gradual evolution in the use of the various primary sources of energy in the time period from 1780 to 2010 for the USA. Most of the energy we use comes from the Sun either directly or indirectly. However, the discovery of nuclear energy in the mid-twentieth century signaled yet another change in the relation between man and energy: *the access of humanity to energy sources which are independent of the Sun* (see Chap. 7).

From the eighteenth century onward, we have witnessed the ascendancy of modern civilization, grounded on science, technology and increased consumption of energy. Critical in this period were the scientific and technological advances concerning the transformations of energy, rooted in the first and second laws of thermodynamics, the law of conservation of energy and the law of entropy, respectively. According to the first law of thermodynamics, the energy of an isolated system is constant and expresses the ability of the system to produce work. According to the second law of thermodynamics when the energy of a system is transformed (changed) from one form to another, its entropy increases and its ability to do work decreases. Thus, while the first law of thermodynamics states that it is possible, for example, to use heat to generate work, the second law of thermodynamics describes how this can be done efficiently.

Physical science provided the basic understanding of the mechanisms of transforming various forms of energy into other *usable* forms of energy, and gradually accelerated the production of useful energy and the development of efficient ways to transport, store, and use energy. With the aid of science, man, in the last two centuries, built machines to produce work using heat and steam, initially by burning wood and coal. Subsequently, he found oil and used it as fuel in engines; located underground sources of natural gas which he used for fuel, lighting, heating, cooking and electricity generation; discovered electrical energy and safe ways to generate, transport and use it; and learned to produce electrical energy from uranium rocks. In this way, and while still the generation of useful energy relies heavily on thermodynamics, over the last 70 years or so, the science of energy has expanded into the fields of nuclear and plasma physics. This expansion is principally due to Einstein's scientific discovery that mass and energy are equivalent quantities. This equivalence, for the present discussion, can be expressed as

$$\Delta E = \Delta mc^2 \tag{2.1}$$

where c is the speed of light in vacuum and Δm is the quantity of mass which is transformed into energy ΔE. According to Eq. (2.1), enormous amounts of energy can be generated by the transformation of mass into energy. Modern science has shown this to be possible in two fundamental ways: *nuclear fission and nuclear fusion*. Physics taught man how to "burn" the fissionable elements of uranium ($^{92}U_{235}$ and $^{92}U_{233}$) and plutonium ($^{94}Pu_{239}$), and how to create conditions for fusion of the light elements, for instance, the isotopes deuterium (D) and tritium (T) of hydrogen and generate energy (see further discussion in Chap. 7).

Modern civilization, and the "affluent society" it sustains, is largely identified with industrial development based on science and technology and abundant energy. The basic criterion of progress of modern civilization has been and remains the *continuous growth, the continuous increase of the Gross Domestic Product (GDP). In pursue of this goal, modern civilization has largely ignored the natural limits of the energy sources and the impact of energy production and use on the Earth.*

In the past, the amount of energy necessary to support civilization was small, and the then known forms of usable energy were few. Today, civilization needs incredible amounts of energy in vastly varied forms. The standard of living of every nation is rooted in energy and no other nation exemplifies this fact better than the

United States of America. Indeed, the greatest choice to the average citizen in human history is today available in the high-energy-consuming societies of the developed countries of the world. It can in fact be said, that the basis of modern civilization is an incessant flow of energy, which relies heavily on the production and use of electrical energy and its delicate "conditioning" to meet the needs of modern technologies. High-value forms of energy, for instance delicate pulses of electrical energy or of light, are at the heart of modern technology, for example, information technologies. It can, in fact, be said, that every technology is based on the *availability of energy of a particular form,* and that every conceivable technology that could be realized in the future, will be realized, *if there is energy available for it in the particular form required.* And so, the quest for energy marches on unabated. Modern civilization's blind belief in the necessity of continuous development ("progress") engenders dangers for its sustainability and its cherished freedoms. It is certainly a major challenge to humanity.

Prior to addressing the energy needs of the world's underdeveloped countries and the relation of energy to poverty, let us look at a few distinct characteristics of the modern industrial civilization over the last 200 years or so, which tightly relate to energy and increased energy consumption.

2.3 Distinct Characteristics of Modern Civilization Related to Energy

2.3.1 Increase in World Population

For millennia, the increase in the population of the Earth was small[27] (Fig. 2.3a). The rapid increase in production and consumption of energy, and the consequential increase in food production and economic growth over the last two centuries, has caused a precipitous increase in world population (from about 1 billion in 1800 to about 7 billion in 2000) during the so-called energy revolution[28] (Fig. 2.3b). The discovery of new sources of energy, writes historian Fernández-Armesto,[29] almost always caused an increase in human population. According to UN (2011)-estimates[30,31] world population will surge past 9 billion before 2050 and will reach 10 billion by the end of the twenty-first century. Most of this increase will be in the low-income regions of the world as can be seen from Fig. 2.3c (see endnote 30).

The global rates of population increase have exerted enormous pressure on man's social and political systems and so had the resultant new demographics. The challenge therefore is how to limit population growth to secure conditions for a respectable life for every human being. In the past, Malthusian predictions of apocalyptic collapses have mostly been averted by the advent of cheap energy, the rise of science and technology, and the green revolution. However, consumption of resources such as food, water, fossil fuels, timber, etc. has grown enormously in the developed world, and perverse subsidies encouraged overuse of resources.

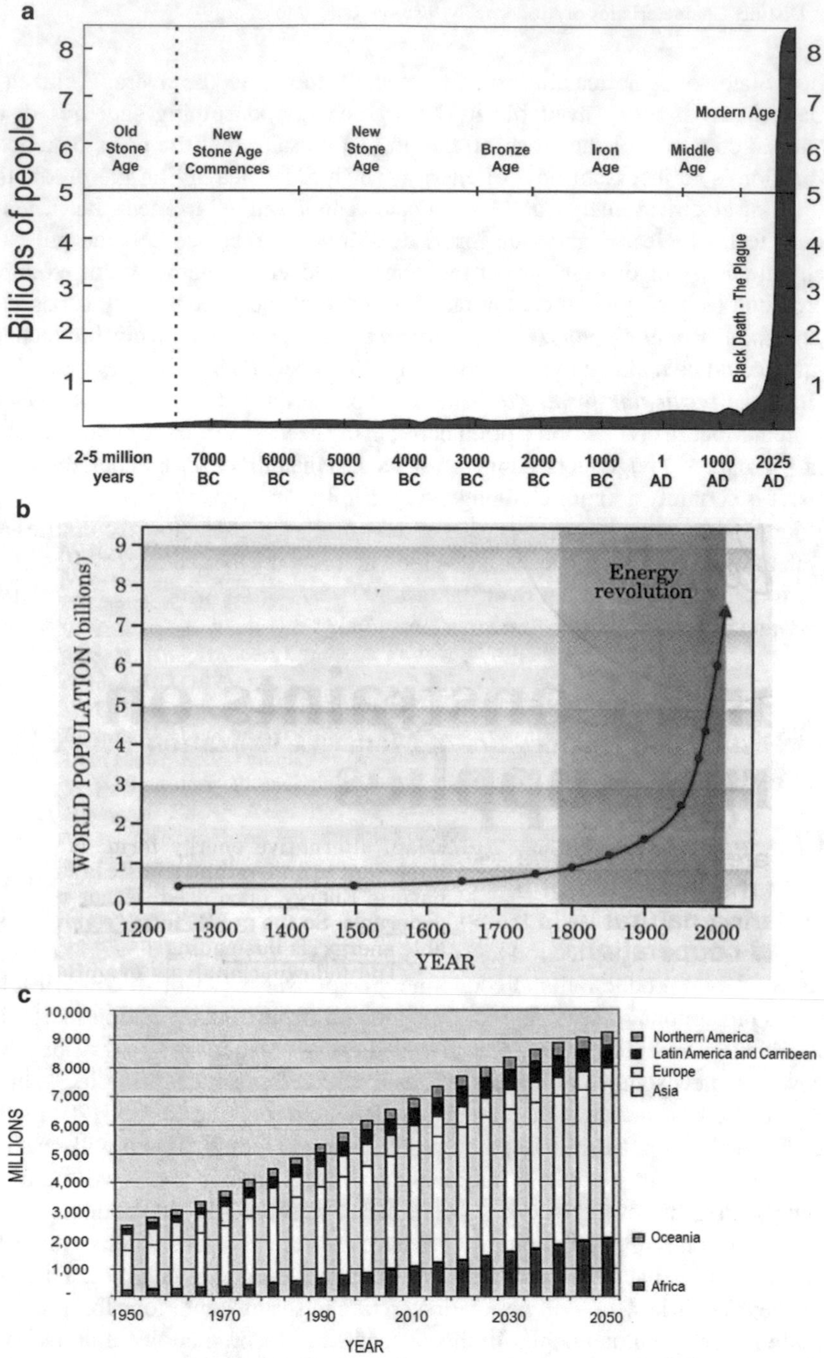

Fig. 2.3 (**a**) Increase in the population of the Earth through history (see endnote 27) (**b**) World population growth since the thirteenth century and the precipitous increase in the population of the Earth in the last two centuries because of increased energy use. (Reproduced from Paul B. Weisz (see endnote 28) with the permission of the American Institute of Physics) (**c**) World population to 2050 – UN data, for different regions of the world (see endnote 30). (Data as plotted by the US DOE)

2.3.2 *Increase in Urban Population*

Civilization, it is argued,[32,33] revolves around cities, and cities are not new; Chandler[34] notes that Rome around 100 AD and Cordova around 1000 AD each had 450,000 citizens. At the end of the eighteenth century, however, the industrial revolution caused a sharp increase in population and urbanization. The use of energy in ever-increasing quantities, new technologies, new industries, new materials, new products, new jobs and higher incomes[35,36] led to a flow of large numbers of people from the rural and agricultural areas into the city. Because of increased energy consumption, the cities grew bigger and the number of big cities increased precipitously (Table 2.2), some becoming megacities and megalopolises. Indeed, not just the size, but also the shape of the cities has been shaped by energy. This trend is continuing at a fast rate, especially in the developing countries.

It has been estimated by the United Nations (see endnote 33) that in 2007, for the first time in history, about half of the world's population (and about 70% of the world's children) were living in urban areas. Most of the growth in urban population observed in the past few decades, and that which is projected to occur by 2030, has been and would likely be in small and medium-size towns and cities of fewer than one (1) million inhabitants.

The efficient functioning of the technological infrastructure which supports the complex life of the growing city crucially depends on energy; it continuously demands more energy. The increase of the complexity of modern cities and their energy needs make society fragile. The quantity and quality of energy demanded by the combination of high population, transportation and industrial density, many have noted, makes it difficult to power the city by small-scale, decentralized soft energy sources. There is, thus, a need to curb the urban energy greed.

Table 2.2 Increase in urban population (Increase of the number of big cities and their size during the last two centuries (see endnotes 18, 34))[a,b]

Cities with more than one million inhabitants		Cities with more than ten million inhabitants		Percentage of the Earth's population living in cities	
Year	Number	Year	Number	Year	(%)
1800	**1** (4)[c]			1800	< 3
1900	**13** (16)[c]	1900	0	1900	~ 10
2000	**375** (299)[d]	2003	20	2000	~ 50
2010	> **472**[d]	2010	> 26		

[a]See also Scientific American (September 2011) and a Report on "Cities in a globalizing world Global report on human settlements 2001" by the United Nations Centre for Human Settlements
[b]See, also, Reference 20, p. 209
[c]http://en.wikipedia.org/wiki/Historical_urban_community_sizes
[d]http://www.citypopulation.de/world/Agglomerations.html

2.3.3 Increase in the Consumption of Energy, Principally Fossil Fuels

Energy consumption has risen continuously since the start of the industrial revolution. Today, society needs enormous amounts of energy; in the past 50 years, world energy production has increased more than fourfold. In the year 2000, the annual total world energy consumption exceeded 400 billion Joules (400 EJ) (see endnote 18).[37,38] The world energy needs are anticipated to reach 623 EJ by 2035, that is, to be ~ 55% higher than in 2000.[39,40] Energy consumption will continue to grow despite efforts to increase energy efficiency and energy conservation and it is feared[41,42,43,44] that it will deplete the Earth's known energy sources, especially fossil fuels, despite views to the contrary. One thus wonders if such a rate of energy consumption is sustainable and recognizes the challenge of the limits of natural energy sources, especially when the energy needs of the entire humanity are considered.

Thus, while in the past the consumption of energy by society was minor compared with the energy available, the recent increases in energy consumption deplete the Earth's known reserves of fossil fuels at a fast rate. Perhaps new sources of energy will come along to satisfy humanity's energy needs. Until that happens, we need to be cognizant of the fact that in the past, civilizations declined and reversed to a more primitive way of life (whether in India, China, or the Middle East) when they failed to control the consequences of energy demands that exceeded the limits of the available energy.[45,46]

2.3.4 Increase in Resource Consumption and the Consequential Increase in Environmental Pollution and Climate Change

Energy is the most crucial element in understanding the impact of "human footprint" on the environment and climate change. The main sources of energy which have sustained the activities and the infrastructure of modern civilization over the past two centuries – coal, oil, natural gas, and uranium – have serious environmental consequences, and the fossil fuels signal dangerous climate change. The increase in energy consumption affected adversely the Earth.

Two-thirds of the greenhouse gases in the atmosphere originate from the production and use of energy and are due mainly to the fact that over 80% of the world energy consumption comes from fossil fuels. Energy and the environment are strongly coupled, multiply connected, and mutually affected. In the last 200 years or so, the production of energy, the transformation of energy from one form to another, the transport and the use of energy, have affected the environment and have contributed to climate change more than any other single factor in human history. This is principally the reason why in the future everything that relates to energy will be crucially determined, and limited, by its environmental and climate impact. Anthropogenically-driven climate change arising from society's consumption patterns is one of the major challenges of the twenty-first century.

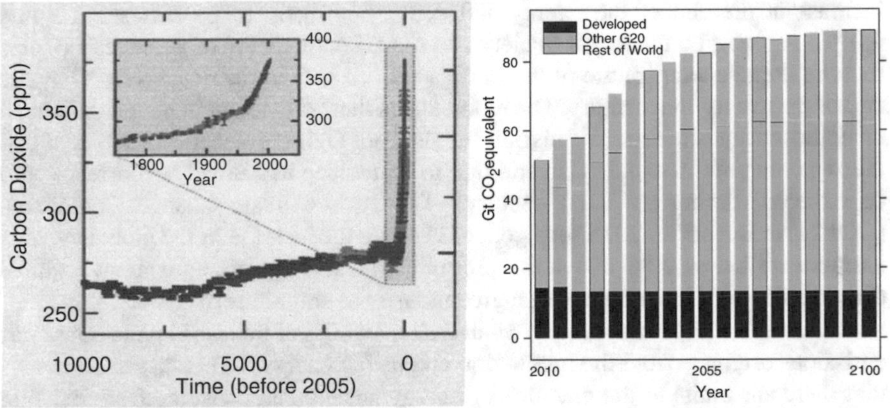

Fig. 2.4 **(Left):** Concentration of CO_2 (in parts per million, ppm) in the atmosphere during the last ten millennia (inset: from1750) (see endnote 47). **(Right):** Current and projected future greenhouse gas emissions for developed countries (dark blue color), other G20 countries (green color), and the rest of the world (light blue color) in this century (see endnote 50)

Figure 2.4 (left) shows the well-known IPCC (Intergovernmental Panel on Climate Change) scientific data on the concentration of the key greenhouse gas, CO_2, in the atmosphere over the last 10,000 years and, in the inset, over the last 250 years.[47] Up to the beginning of the eighteenth century, the concentration of CO_2 in the atmosphere was basically constant (about 280 ppm, that is, 280 molecules of CO_2 in one million of molecules of air), and it was principally originated from natural sources. Since then, the concentration of CO_2 in the atmosphere increased substantially (in May 2013, CO_2 levels in the atmosphere exceeded 400 ppm[48]) and this increase comes from anthropogenic sources, principally energy-related. The same holds true for the other greenhouse gases, e.g., CH_4 and N_2O, which are more potent, have Global Warming Potentials higher[49] than that of CO_2.

Figure 2.4 (right) shows the current and projected future greenhouse gas emissions for developed countries (dark blue color), other G20 countries (green color) and the rest of the world (light blue color) in this century.[50] While emissions from the developed countries are predicted to be reduced between 2010 and 2100, those from the other G20 countries and from the rest of the world are predicted to increase in this time period.

There is convergence of opinion in the scientific community that the average temperature of the Earth's surface increases, although there is divergence of opinion regarding the size of this increase, which could be attributed to deficiencies of the models used in such predictions.[51] According to a policy report by the European Academies Science Advisory Council (EASAC) entitled "Trends in extreme weather events in Europe: implications for national and European Union adaptation strategies",[52] during the past 50 years the global mean temperature at the Earth's surface has increased by about 0.7 °C. While there is scientific consensus that the Earth is warming up because of human actions, there seems to be no scientific consensus as to how hot the surface of the Earth will get and how it will respond. However, there have been estimates predicting (see endnotes 48, 51) that the increase at the end of the twenty-first century is likely to be between 1.8 and 4.0 °C. A report by the Royal Society and the USA Academy of Sciences[53] doubts that the average temperature of the Earth's surface will be restricted to 2 °C by the end of the twenty-first century. The report states that "If there were no technological or policy changes to reduce emission trends from their current trajectory, then further warming of 2.6 to 4.8 °C in addition to that which has already occurred would be expected during the 21rst century." The Paris Climate Change Conference (COP21) in December 2015 set a goal of keeping the increase in the global average temperature below 2 °C above the pre-industrial levels. This agreement requires drastic reductions in anthropogenic greenhouse gas emissions by 2050.

It is estimated (see endnote 27)[54] that well over 60% of the world greenhouse gas emissions originate from the production and use of energy.[55] This fact alone shows that there are limits to the amounts of energy society can consume from existing energy sources, and difficulties in sustaining modern civilization, especially since until recently high energy consumption involved mainly only a part of the world. It must be emphasized that these consequences will be exacerbated in the future because the consumption of energy ***is*** expected to continue increasing, largely because of the energy needs of the developing countries.

In the twenty-first century the environmental challenges are predicted to be immense due to the needs of a projected total population of ~10 billion and the growing energy demands. A transition to alternative less polluting energy sources is therefore needed, especially in developed countries and newly industrialized nations particularly India and China.

The recent data (Fig. 2.5) referring to China[56] make this point clear. Because of the increased consumption of energy, the Gross Domestic Product (GDP) of China has been increasing by almost 10% annually for 30 years or so (bottom, **blue** color

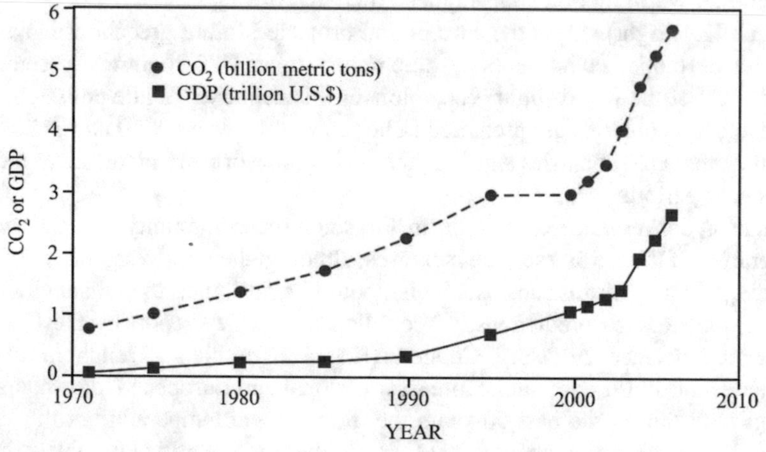

Fig. 2.5 The GDP of China has increased substantially over the time period shown in the figure, but a parallel increase has resulted in the emissions of CO_2 from fuel combustion (see endnote 56)

curve). As a direct consequence of this, the CO_2 emissions in China over the same period have increased alarmingly (top, **red** color curve). The CO_2 gas emissions in China in 2007 have surpassed those in the USA, and in 2010 they accounted for about 24% of the total world CO_2 gas emissions.[57,58]

Especially serious are the environmental and climate-change consequences of the burning of enormous quantities of coal by China.[59] Coal-fired electricity generators in China, in 2011, represented 78% of the 1 billion kilowatts of installed capacity (see endnote 58). Without developing other types of generating capacity, such as nuclear and renewables, China's demand for coal will likely account for more than half of the world's total demand for coal (see endnote 58).[60,61] According to Liu and Diamond (see endnote 56) environmental pollution and deterioration in China are such that two thirds of China's cities suffer from lack of clean water, and according to Liu and Yang[62] "two-thirds of China's 669 cities have water shortages, more than 40% of its rivers are severely polluted, 80% of its lakes suffer from eutrophication, and about 300 million residents lack access to safe drinking water".

Climate change and environmental impact of the production and use of energy began first in Europe and then in North America, that is, long before the recent industrial development of the underdeveloped countries of the world. Clearly, a transition to alternative energy sources in developed countries and China (and India) is needed and should begin by reducing the burning of coal.[63] It is noted also that China is investing heavily in "cleaner" technology, renewable energy sources and nuclear power and that China's growth is expected to become less energy intensive in the years ahead. By 2035 China will be using a fifth of all global energy, a 75% increase over 2008.[64,65,66]

China, writes Ferguson,[67] "remains a poor country with 150 million of its citizens – nearly one in ten – living on the equivalent of $1.50 a day or less. Inequality has risen steeply in China: an estimated 0.4% of Chinese households currently own ~70% of the country's wealth." As mentioned earlier in this section, there are chronic problems of air, water and ground pollution; according to Science magazine,[68] China's environmental ministry acknowledges "cancer villages", that is, "locations where exposure to environmental hazards, often water pollution, is believed to have contributed to elevated cancer rates."

Measurements of the global average temperature, the global average sea level, and the Northern Hemisphere snow cover show that the Earth is being heated up, the sea level is rising, and the snow-cover of the Northern Hemisphere is melting. The rise in the global average temperature causes a series of other extreme climate changes and may be responsible for the observed increase over the last half century in the number of extreme atmospheric phenomena, for instance floods (see endnote 18).[69,70,71]

Recently, increased attention has been drawn to the occurrence of extreme weather events (see endnote 52).[72] An EASAC report on the subject states (see endnote 52) that "weather-related catastrophes recorded worldwide have increased from an annual average of 335 events from 1980 to 1989, to 545 events in the 1990s, to 716 events for 2001-2011". Interestingly, it is asserted in the EASAC report that "it is not primarily the change in the mean climate variables such as temperature,

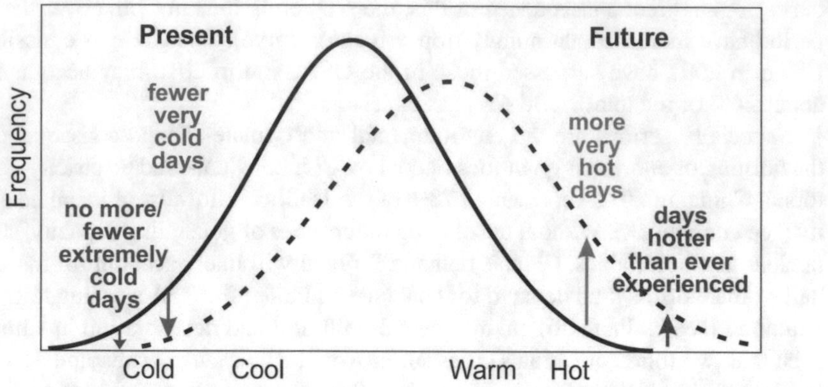

Fig. 2.6 Schematic illustration; increasing the mean temperature shifts the solid curve to the dashed curve and results in an increase of the frequency of record hot days and leads to days hotter than has ever been experienced before (see endnote 52)

precipitation, or wind, or in derived variables like storm surge or water runoff, but rather the changes in the extremes of these variables that pose serious risks". There is thus "a need to recognize the effects of a change in the average value for a weather event (e.g., temperature or precipitation) on the future frequency and intensity of that event". This is illustrated schematically in Fig. 2.6 for the case of temperature: increasing the mean temperature shifts the solid curve to the dashed curve and results in an increase of the frequency of record hot days and leads to days hotter than has ever been experienced before, even though extremes have small probabilities.

Extreme weather events (heat waves, floods, droughts, storms) recur irregularly and can have economic and social impacts. The reduction of the factors driving climate change (mitigation) has the benefit of reducing the cost of adaptation (see endnote 52).

2.3.5 *The Increased "Negative" Use of Energy (Increased Use of Energy for Destruction and War)*

The availability and use of enormous amounts of energy advance civilization, but also, often, they disrupt civilization and hinder its advancement. Time and again in history, nations used the energy they possessed to conquer other nations and enslave other peoples. In wars between nations prevailed as a rule those nations which had in their disposal more energy, and this holds true also for World War II (see endnote 20).[73]

Today, as in the past, elevated levels of energy consumption are a prerequisite of military might and superiority, a motivation for assumption of offensive wars, and a means for political power and supremacy. Ultimately, the nation which controls the largest energy resources will likely prevail. Even countries which do not have high-level science and advanced technology but are in possession of large energy resources enjoy disproportionate international influence. Countries, such as the USA, achieved and maintain their enormous power and world influence largely by the massive amounts of energy they use.

The technologies of energy and the technologies of war are strongly coupled and multiply connected. Man, uses energy to create civilization, but he, also, time and again, threatens civilization by using energy for destructive purposes.

2.3.6 *Increase of Societal Complexity*

It has been argued (see endnotes 45, 46)[74] that the more energy a certain society uses, the higher is its complexity. Thus, when the amount of energy a society has at its disposal and uses is reduced, or when the amount of energy a society needs stops increasing while its energy needs continue to grow, that society is at risk if it fails to secure new sources of energy, or if it is unable to use more efficiently the energy it has. Many historians believe that a significant common reason for the collapse of past civilizations was their weakening by over exploitation of their energy resources, which made them vulnerable and ultimately unsustainable (see endnotes 18, 20, 45, 46, 74).[75]

For how long then will modern civilization continue to increase its energy consumption and complexity before it becomes unstable and hence non-sustainable? If the energy consumption and the complexity of today's society are the determining factors which define societal progress, for how long can these factors continue to grow before today's technological society becomes unstable? Complex systems which have no other choice but to go on increasing in complexity and growth eventually collapse.

It, thus, seems that the continued increase of complexity of modern civilization makes it fragile and operationally vulnerable. Modern civilization crucially depends on the smooth operation of a multitude of extremely complex interdependent and interconnected systems, and *each one of these systems depends on energy*. It suffices the malfunctioning of some of the components of the system to trigger a cataclysmic failure in the operation of the entire system, causing its collapse. *One is thus forced to conclude that the sustainability of modern civilization is in danger without huge new amounts of energy*. Otherwise, civilization as we know it is unstable, because too many of its functions are increase-only; collapse occurs when the acquisition of energy stops continuing to grow, but not the need for it, and adaptation evades a mismanaged society.

2.3.7 *Increased Gap in the Standard of Living Between the Energy Rich and the Energy Poor Peoples*

It is well recognized that there is an asymmetry in the consumption of energy between the peoples of the world, which separates them into poor and rich. The widening gap between the rich peoples and the poor peoples of the world largely reflects the existing difference in their energy consumption levels and the lack of access to modern energy services by the latter. *Peoples' poverty is in fact peoples' energy poverty*. The industrialized nations rely on abundant fossil fuels and electricity for their standard of living, while the poor regions of the Earth (especially their rural areas) rely, as a rule, on biomass and animal waste. The enormous energy consumption in the advanced countries impacts heavily on the energy poverty of the developing countries and their ability to cope with the consequences of climate change.

Perhaps one way the relation between the consumption of electrical energy and the standard of living among the peoples of the world could be illustrated is by the way lighting is distributed across the surface of the Earth at night, as is shown[76] in Fig. 2.7. Although the map provides a beautiful composite view of the patterns of human settlements across the surface of the Earth, clearly there are settlements without sufficient lights to be seen on the map. There are, for instance, only a few regions with abundant lighting in Africa! A large part of humanity still today lives in the dark and is isolated.

It remains an undeniable fact that almost every problem of the poor countries of the world is related directly or indirectly with energy. Therefore, to wipe out poverty and to secure decent life for the billions of people in the poor countries, we need to wipe out the energy poverty of these countries, and this requires more energy consumption in the future.

Energy can be an effective weapon against poverty. Yet one wonders. If the recent increases in the use of fossil fuels by the developing countries, foremost of Asia (China and India), have had such negative environmental consequences and

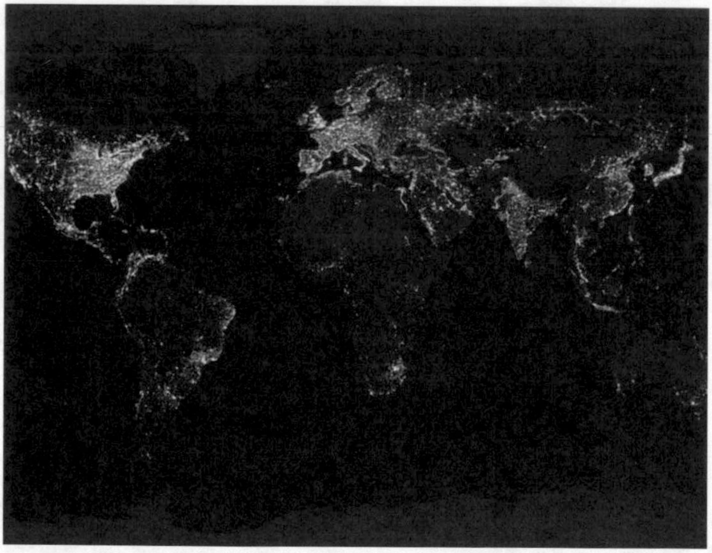

Fig. 2.7 Distribution of lighting across the surface of the Earth (see endnote 76)

have resulted in rather marginal improvements in the standard of living for many of the people of those countries, then, pessimistically, one is led to conclude that the "health" of the Earth will continue to be in danger and the survival and dignity of a large part of humanity will fall to still lower levels without the development of new energy sources, economically feasible and environmentally friendly. *Without abundant clean energy, neither the environment nor poverty can be dealt with effectively*.

The dimensions of energy poverty are vast and their consequences serious, often devastating. The world's poor need energy, foremost modern forms of energy, and especially *affordable electricity*. Contrary to the case of the developed countries (see Chap. 7), electrical energy is not available (and not affordable) in sufficient quantities to the poor regions of the Earth. In Fig. 2.8 is plotted the share of population with an income of less than $2 per day and electricity consumption for Sub-Saharan Africa, North Africa, Developing Asia and Latin America.[77] The data show that countries with a large fraction of their population having an income of less than $2 per day tend to have low electrification rates.

The fundamental question, therefore, is how the enormous energy needs of the developing countries can be effectively satisfied, without endangering further the Earth and the sustainability of civilization. This challenge necessitates more clean energy and responsible management of all energy resources (See further discussion on energy and poverty in Chap. 7).

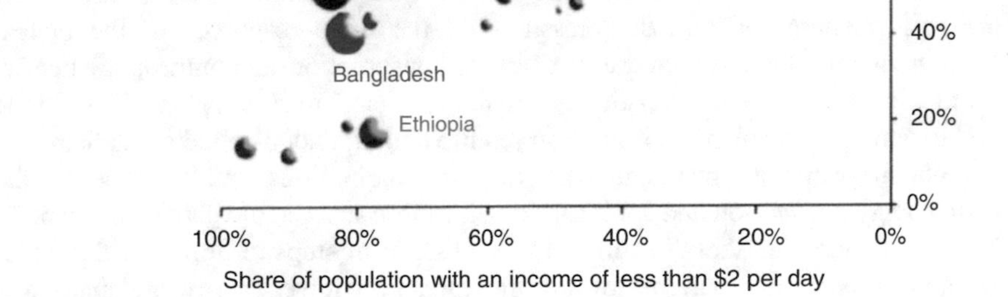

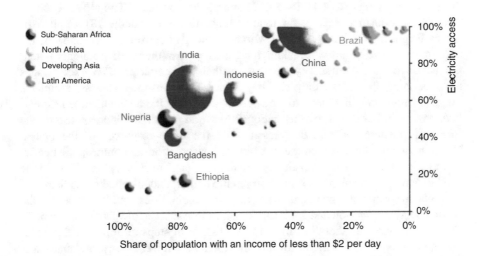

Fig. 2.8 Share of population with an income of less than $2 per day as a function of access to electricity (The size of the bubble is proportional to the country's population) (see endnote 77). (IEA Publishing. License: www.iea.org/t&c)

2.3.8 *Resource Consumption and Sustainability of Modern Civilization*

In this Section, we look briefly at resource consumption and the role of science and science-based technology in the sustainability of modern civilization referring specifically to two aspects of resources: (i) the balance between ***availability and consumption of resources*** and (ii) the interconnection of ***energy, food, and water***.

2.3.8.1 Tame Consumerism

A crucial element for the sustainability of modern civilization is the balance between availability and consumption of resources. It is unlikely that technology alone will be sufficient for society to achieve this balance; society *must tame consumerism through cultural change and adaptation.*

After WWII, the consumer society especially in the USA became a phenomenon of the "masses" and this phenomenon continues unabated (consumer credit to buy everything: homes, appliances, telephones, washing machines, air conditioners, clothes dryers, TV sets, microwave ovens, mobile phones, internet access and so on); if anything has changed since, it is the intensity of the phenomenon and its world-wide expansion.

Global prosperity has had a long record of providing more goods and services and it has been a driver of global production and consumption. Over the past 50 years there has been[78] a 5-fold increase in global economic growth, which is greater than the increases in energy (4.4), food (2.7) and population (2.2). The top 20% of the world's rich consume 85% of the total commodities (see endnote 78). Capitalists, writes Ferguson,[79] understood "that workers were also consumers.... (and), as the case of the United States was making increasingly clear, there was no bigger potential market for most capitalist enterprises than their own employees." Worker's wages must be kept up so that they can buy more! The consumer society, maintains Ferguson, posed a lethal threat to the Soviet system. Will it, and the untamed capitalism, pose a similar threat to other nations? As world population increases so do the forms of *consumption* and the pressures on the finite resources of the planet. Consumerism drives the increase in resource consumption and anthropogenically-driven climate change and constitutes a major challenge to *development.* Would the need to tame uncontrolled consumption require taming uncontrolled capitalism?

Technology and energy resources in various combinations could, in time, put the world energy production and consumption on a more sustainable path (see Chap. 7). However, they are unlikely to suffice in the absence of steps to limit world population, shift consumption patterns towards those that have lower energy- and materials-intensity and reduce fossil fuel consumption by building a low-carbon economy. The immensity of environmental and climate change challenges is paralleled by that of food and energy security, water scarcity, ecosystem protection, and the problems of urbanization. *Curb the increase in resource consumption and conserve resources, and start with energy.*

2.3.8.2 Conserve Energy

No state of the future can be envisioned without first envisioning the future state of energy resources. Key challenges are: (i) secure the supply of reliable and affordable energy, (ii) effect a rapid transition to a low-carbon, efficient and environmentally friendly system of energy supply, (iii) develop innovative ways to access existing forms of energy, and (iv) search for new sources of energy and new forms of energy that can be transformed to useful energy sources.

The subject of energy as such will be discussed in Chap. 7 and for this reason we restrict ourselves here to only a few facts relevant to resource sustainability and energy conservation, energy efficiency and energy intensity. Sustainable energy is not just using renewable energy or shifting from combustion economy to solar electric economy; it is **also** about conserving energy, using energy wisely and efficiently, reducing energy intensity, as well as increasing recycling (*cyclic economy*).

Let us look at *energy conservation*. Energy conservation is the cleanest of all energy sources and a significant source of energy today and in the future (see endnote 18).[80] It is, also, an effective way to reduce greenhouse gas emissions; it can be looked at as an inexhaustible source of energy that depends on the responsible behavior of every citizen and every county. There are countless ways that each one of us can save energy. Figure 2.9[81] shows two of them. The **blue curve** gives data on the consumption of electrical energy by refrigerators in the USA (also in other counties). Since 1974 – when the cost of energy began to increase and measures for energy conservation began to be implemented – improvements in the insulation and refrigerant systems using new materials and technology resulted in significant reductions in energy consumption by these appliances. On the same figure, the

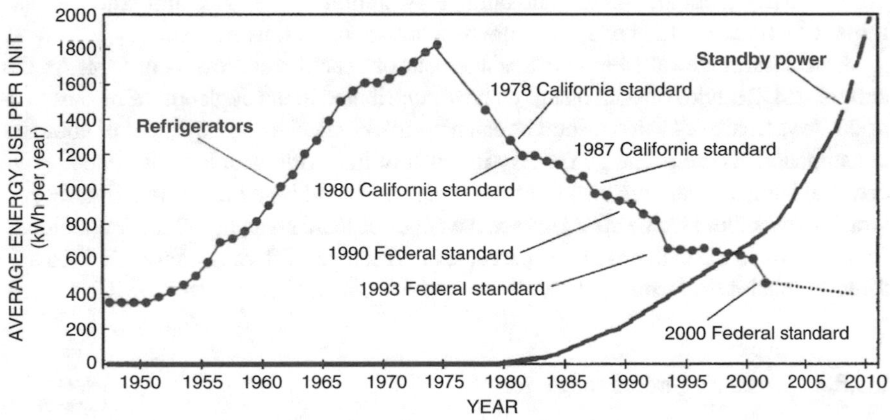

Fig. 2.9 Consumption of electrical energy by refrigerators in the USA (blue-color curve), and consumption of electrical energy by electronic devices in a standby mode (red-color curve). (Reproduced from Leon R. Glicksman (see endnote 81) with the permission of the American Institute of Physics)

red curve refers to the "standby" consumption of electrical energy by electronic equipment, mainly computers. Electronic equipment consumes in its totality enormous amounts of energy even when it is not used, but it is "on" and "inactive" or in "standby" mode. The consumption can be remedied with changes in the appliances themselves, and, of course, by switching off electronic devices when not in use.

Attention should be drawn also to a similar recent problem: the enormous energy consumption of present and future computer servers and data centers. Meijer[82] notes that soon after the Internet took off in the mid-1990's, data centers became common place; information technology industry needs data centers. The energy consumption challenges posed by such centers are considerable, due to the power consumption of the cooling infrastructure that is required to keep the microelectronic components from overheating (the transistors inside the microprocessors). In 2009, an estimated 330 terawatt-hours of energy (about 2% of the global electricity production) was consumed to operate data centers worldwide[83] with considerable economic and ecological consequences. Meijer points out that the information technology industry needs "efficacious concepts to reduce the energy consumption of data centers." He notes that, since the electrical energy which is supplied to the computer system is eventually entirely converted into thermal energy, "liquid cooling and deploying waste heat appear to become imperative in the drive for improving the data-center energy efficiency".

Besides the technical aspects of the problems of energy and the environment, society's approaches and responses toward energy and the environment are also crucial. To choose wisely the most efficient and environmentally friendly energy source, or to optimize a combination of them, it is necessary to consider *every energy source* that is available. It is also necessary to consider adjustments in long-held lifestyles. For instance, most people, certainly those in the developed world, regard mobility as a fundamental freedom. Yet, this freedom relies on a transport system which is a massive consumer of fossil fuels and consequently impacts seriously and negatively on the environment and on climate change. Adoption of lifestyles and land use patterns, which reduce the need for motorized transport are thus needed.

A successful stewardship and management of shared resources is crucial. As the actions and lifestyles of each one of us have contributed to the depletion of resources, a good stewardship of shared resources can help ameliorate the consequences of untamed consumption. There is energy poverty in much of the world which feeds general poverty and hunger. There are also rights to energy and food based on human dignity and worth. Why should we be afraid to face the ethics of these challenges? If the conditions for a just world mean that we must perceive our resources differently, produce food differently, inhabit the terrain differently, see our freedom differently, let it be it.

2.3.8.3 Increase Food Production

We referred to this topic earlier in this Chapter but we believe we should revisit it here again because it is a distinct characteristic of modern civilization strongly dependent on science-based technology and the needs and values of society.

Food production is vitally dependent on water, another natural resource which we take for granted, and often abuse, forgetful of the fact that about 1 billion people do not have access to clean water and consequently over 3.5 million people die each year from water-related diseases.[84,85] The EU Joint Research Center reports (see endnote 84) that 18% of the world's land area is used for agriculture and accounts for about 85% of global freshwater use (irrigated crop yields are globally 2.7 times higher than those of rain-fed farming).

Continuing population and consumption growth will mean that the global demand for food will increase (perhaps double by 2050). Growing competition for land, water and energy, as well as overexploitation of fisheries and biofuel policies are among the factors that would affect the ability to produce food in the future. Recent studies (see endnote 17) claim that roughly 30% to 40% of food in both the developed and developing world is lost to waste. In the developing world losses are mainly attributable to the absence of food-chain infrastructure and the lack of knowledge or investment in storage technologies on the farm. In India, for instance, it is estimated that 35–40% of fresh produce is lost because neither wholesale nor retail outlets have cold storage (see endnote 17). By contrast, in the developed countries, pre-retail losses are much lower; however, those arising at the retail are high; food service and home stages of the food chain have grown dramatically recently. There is thus a need for changes in the way food is produced, stored, processed, distributed and accessed, and for modern technologies that would enable production and distribution of more and better food (e.g., food-chain infrastructure and storage technologies).

According to Uma (see endnote 12) there are at least one billion poor people living with chronic undernourishment and the United Nations Millennium Development Goal of substantially reducing the world's hungry by 2015 will not be met. Uma writes "The main battlegrounds for poverty reduction are Asia and Africa, where 97% of the world's food-insecure reside…. Lifting a billion people out of poverty and feeding an extra 2.3 billion by 2050 will require increasing cereal production by 70%, doubling the output of developing countries. In Sub-Saharan Africa where more 'ultra-poor' live, developing technologies to boost productivity is especially difficult because of greater threats from pests and diseases, poorer soil, and drought". This is one of the reasons why Genetically Modified Foods (GMFs) are important.

In the fight against world hunger, two areas are of paramount significance: energy and biotechnology/genome technology.[86] Energy is the basis of food supply. Farmers have become more energy incentive and the industrial food system depends on the same sources of energy as electricity and transport, that is, on fossil fuels. The role of science and science-based technology in increasing food supply can hardly be overstated, especially biotechnology and genomics. Genetic modification enables the transfer of genes artificially from one organism to another for a specific purpose (e.g., to increase yield and nutritional value, protect against pests and diseases, enhance survival in hostile environments). The speed and costs at which genomes can be sequenced or re-sequenced today means that these techniques can be more easily applied to develop varieties of crop species that will yield well in challenging environments. Currently, the following GM crops are among those which have been commercialized: maize, soya bean, cotton, canola, squash, papaya, sugar beet, tomato, sweet pepper and alfalfa. They are grown primarily in North and South America, South and East Asia. Crops such as sorghum, millet, cassava, and bananas, are stable foods for many of the world's poorest communities. In the future, options involving highly polygenic traits may become feasible.

In efforts to boost agricultural productivity in the world's poor regions, attention has been drawn to Africa (see endnote 13).[87] Africa must embrace technologies that enable it to produce more and better food with less effort. By 2011 only a few African countries could grow genetically modified crops, partly because of restrictive national biosafety policies that impose regulatory barriers to the adoption of agricultural biotechnology. African continent commercialized GM crops include maize, cotton and soya beans. Maize is the most widely grown stable crop in Africa and it provides food for more than 300 million people. Africa needs GM disease-resistant crops.

Yet many critics question the role of biotechnology in world agriculture; like other agricultural practices, biotechnology is not risk free. Coexisting with the benefits of genetic modifications of plants and animals are known and unknown health risks and food safety (allergies, toxicities, nutrient imbalances, decreasing diet diversity); concerns about possible adverse impact on the environment such as the transfer of GM genes to wild relatives and the development of resistance to pests that need to be taken seriously and to be kept under constant review; socio-economic and ethical issues such as control of agricultural biotechnologies and intellectual property rights (see endnote 14).[88] While the production of a GM crop passes through many stages of stringent scrutiny, there remains skepticism over their safety to humans and the environment despite assurances to the contrary.[89,90]

As mentioned earlier in this Chapter, the issue of GM crops still divides the EU and many believe that it might be an issue of public trust and public fears not unlike other agricultural practices. Many scientists believe that this discussion has become highly politicized and polarized in some countries, particularly in Europe; it is argued that Europe's lack of trust in GMOs may reflect a wider distrust of science. Similar attitudes prevail concerning shale gas and nuclear power. It seems that despite the introduction of rigorous science-based risk assessment of GMOs,[91] the European Union differs from most of the world in its opposition to the use of genetic modification in agriculture. There is thus a need for this technology to gain greater public acceptance and trust so that it can take its place as one among a set of technologies that improves global food security. An EASAC report on the subject states (see endnote 14) "Our view is that genetic modification is a potentially valuable technology whose advantages and disadvantages need to be considered rigorously, on an evidential, inclusive, case-by-case basis. Genetic modification should neither be privileged nor automatically dismissed."

It should be noted however, that opposition to and fears about GM foods is not restricted to Europe. There is, for instance, widespread fear among the Chinese people that GM food is unsafe to eat.[92] According to Wang (see endnote 92) while grain in China "nearly doubled between 1978 and 2013, the increase was driven by a six-fold rise in the use of chemical fertilizers, which pollute the land and the water". Wang maintains that Chinese GM technology has the potential to produce more food with less pollution, but cultivation of GM crops, including rice, is banned. Clearly the public's concerns about GM food safety must be addressed.

Energy and food production have encountered each other in a particularly important way in the production of biofuels.[93] The conversion of biomass from various sources (including waste) may, in conjunction with other energy sources, help to make society less dependent on oil. Although biofuels may burn clean, their production still yields CO_2 and, in some instances, it is in competition with food. The food vs. fuel debate is a genuine issue as the rapidly increasing demand for biofuels can substantially distort global food markets. Any use of GM technologies for production of biofuels, especially when it involves large-scale cultivation of monocultures, would likely connect with long-standing concerns, particularly in Europe, over the use of this technology. The issue of biofuels raises several other concerns[94,95] and requires an informed discussion.

The crucial role of scientific research and science-based technology to enhance food production, achieve food security and solve possible new problems whether to human health or to the environment related to food production, can hardly be overstated. Promising innovative approaches include:

- Those in the energy sector as is applied to food and water for all three phases of energy utilization: energy conversion, energy transfer and energy storage. Especially significant is the role of renewable energy.
- Scientific research to identify plants which fix water from the air even in the driest air of deserts. Possibly, also, raise plants with seawater irrigation of desert lands.
- Take full benefit of the advances in molecular biology and biotechnology for food production and explore options involving highly polygenic traits.

In the beginning of this Chapter we mentioned that in the past the reasons for societal collapse were largely because resources were overexploited. We end this Chapter by stressing the reasons for societal success: *The change in human behavior based on human values and grounded in science and science-based technology*. Then and only then, we can successfully face ***the ultimate challenge of modern civilization, the protection of humanity and the respect for human dignity.***

References and Notes

1. Loucas G. Christophorou, *Science-Based Technology and Society,* keynote lecture, Proceedings of the International Conference "Technology + Society = ? Future", Momir Durović (Ed.), Podgorica 2016, pp. 27–38.
2. It is conjectured (Reference 3, pp. 152–157) that cancer fundamentally involves mutations in four or more genes, and that the fact that it takes a sequence of four or more defective genes to cause cancer probably explains why it often kills years after an original incident (e.g., radiation exposure).
3. Michio Kaku, *Physics of the Future*, Anchor Books, New York 2011.

4. Nicolas Wade, The New York Times, March 19, 2015.
5. European Academies Science Advisory Council (EASAC), *Synthetic Biology: An Introduction*, January 2011.
6. R. Breitling, E. Takano and T. S. Gardner, Science **347**, 9 January 2015, Editorial, p. 107.
7. M. K. Cho and D. A. Relman, Science **329**, 2 July 2010, pp. 38–39.
8. There is an extensive bibliography on these issues (for example, in *Science, Nature, Proceedings of the US National Academy of Sciences*). See, also, D. Sprinzak and M. B. Elowitz, Nature **438**, 24 November 2005, pp. 443–448; Molecular Systems Biology, 2007 EMBO and Nature Publishing Group, Editorial, *Synthetic Biology: Promises and Challenges*, pp. 1–5; M. Schmidt, A. Kelle, A. Ganguli-Mitra and H. de Vriend (Eds.), *Synthetic Biology – The Technoscience and its Societal Consequences,* Springer 2009; E. Parens, J. Johnson, and J. Moses, *Ethical Issues in Synthetic Biology – An Overview of the Debates*, Woodrow Wilson International Center for Scholars, June 2009; R. Kwok, Nature **463**, 21 January 2010, pp. 288–290; M. Elowitz and W. A. Lin, Nature **468**, 16 December 2010, pp. 889–890; B. Erickson, R. Singh and P. Winters, Science **333**, 2 September 2011, pp. 1254–1256; N. Nandagopal and M. B. Elowitz, Science **333**, 2 September 2011, pp.1244–1248; W. C. Ruder, T. Lu and J. J. Collins, Science **333**, 2 September 2011, pp. 1248–1252; S. Joyce et al., *Synthetic Biology*, The National Academies Press, DC, 2013, ISBN: 978-0-309-22583-0; European Academies Science Advisory Council (EASAC), *Realising European Potential in Synthetic Biology: Scientific Opportunities and Good Governance,* December 2010, ISBN: 978-3-8047-2866-0; EASAC, *Synthetic Biology: An Introduction*, January 2011; Volker ter Meulen, Nature **509**, 08 May 2014, p. 135.
9. Nature-Editorial **518**, 19 February 2015, p. 273.
10. F. S. Collins, *The Language of God*, Free Press, New York 2006.
11. H. Pearson, *Genetics: What is a Gene?* Nature **441**, 25 May 2006, pp. 398–401.
12. Uma Lele, Science **326**, 26 March 2010, Editorial.
13. Brian Heap and David Bennett (Eds.), *Insights – Africa's Future... Can Biosciences Contribute?,* Lavenham Press, UK 2013.
14. EASAC policy report 21, *Planting the Future: Opportunities and Challenges for Using Crop Genetic Improvement Technologies for Sustainable Agriculture*, June 2013, ISBN: 978-3-8047-3181-3.
15. Cynthia Kroet 12.06.2014 ENVIRONMENT.
16. http://www.europeanvoice.com/article/juncker-faces-test-on-gm-crops.
17. H. C. J. Godfray et al., Science **327**, 12 February 2010, pp. 812–817.
18. Loucas G. Christophorou, *Energy and Civilization*, Academy of Athens, Athens 2011.
19. Χρύσανθος Χρήστου, *Ο Μύθος του Προμηθέα και ο Ζωγραφικός Διάκοσμος της Ακαδημίας Αθηνών*, Ακαδημία Αθηνών, Αθήνα 2003 (ISBN: 960-404-018-9).
20. Vaclav Smil, *Energy in World History,* Westview Press, Boulder, CO 1994.

21. H. G. Rickover, *Energy Resources and our Future – Remarks*, Energy Bulletin, December 2, 2006; http://energybulletin.net/23151.html.
22. Άπαντα Αριστοτέλους, Ηθικά Νικομάχεια, Εκδόσεις «Ωφελίμου Βιβλίου», Αθήνα 1979, Α, σελ. 425-442; Aristotle, *Nicomachean Ethics,* 1098a, b.
23. The concept of energy in its modern sense was introduced into science by Thomas Young in 1807.
24. http://en.wikipedia.org/wiki/History_of_energy.
25. Crosbie Smith, *The Science of Energy – A Cultural History of Energy Physics in Victorian Britain*, The University of Chicago Press, Chicago 1998, p. 8.
26. Peter A. O'Connor and Culter J. Cleveland, *US Energy Transitions 1780-2010*, Energies 2014, **7,** 7955-7993; doi: 10.3390/en7127955.
27. http://www.susps.org/overview/numbers.html (https://www.linkedin.com/pulse/alert-critically-endangered-species-homo-sapiens-willemijn-heideman).
28. Paul B. Weisz, Physics Today, July 2004, pp. 47–52.
29. Felipe Fernández-Armesto, *Civilizations,* Simon & Schuster, New York 2001, pp. 444–452.
30. Department of Economic and Social Affairs, Population Division (ESA/P/WP.241), *World Population Prospects – The 2015 Revision*, Key Findings and Advance Tables, United Nations, New York 2015 (data as plotted by the US DOE).
31. MIT Joint Program on the Science and Policy of Global Change, 2012 Energy and Climate Outlook. https://www.linkedin.com/pulse/alert-critically-endangered-species-homo-sapiens-willemijn-heideman.
32. P. Hall P, *Cities in Civilization – Culture, Innovation, and Urban Order*, Weidenfeld & Nicholson, London 1998.
33. United Nations, World Urbanization Prospects: The 2003 Revision, New York, United Nations, 2004; David A. Leon, *Cities, Urbanization and Health,* International Journal of Epidemiology, 2008; **37**, 4–8.
34. Tertius Chandler, *Four Thousand Years of Urban Growth: An Historical Census*, St. David's University press, 1987.
35. According to Bloom, Canning and Fink (see endnote 36), the proportion of a country's population living in urban areas is highly correlated with its level of income.
36. D. E. Bloom, D. Canning, and G. Fink, Science **319**, 8 February 2008, pp. 772–775.
37. If this amount of annual total world energy consumption is expressed as a stack of barrels of oil, the height of this stack is more than 200 times larger than the distance between the Earth and the Moon (see endnotes 18,38).
38. Λουκάς Γ. Χριστοφόρου, *Ενέργεια και Πολιτισμός*, Πρακτικά της Ακαδημίας Αθηνών, τ. 85 Α΄, Αθήνα 2010, σελ. 205-227. Loucas G. Christophorou, *Energy and Civilization*, Proceedings of the Academy of Athens, Vol. 85 A′, Athens 2010, pp. 205–227.
39. International Energy Agency (IEA), World Energy Outlook 2011.
40. Heinz Kopetz, Nature **494,** 07 February 2013, pp. 29–31.

41. See, for instance, Vaclav Smil, *Energy Myths and Realities: Bringing Science to the Energy Policy Debate*, The AEI Press, Publisher for the American Enterprise Institute, Washington, DC 2010.
42. InterAcademy Council Report, *Lighting the Way: Toward a Sustainable Energy Future*, October 2007 (ISBN: 978-90-6984-531-9).
43. James Leigh, *A Geopolitical Tsunami: Beyond Oil in World Civilization Clash,* Energy Bulletin, September 2008; (http://www.energybulletin.net/node/46451).
44. M. King Hubbert, *Energy from Fossil Fuels (historical),* Encyclopaedia of Earth; Science **109**, 4 February 1949, pp. 103–109.
45. Joseph A. Tainter, *The Collapse of Complex Societies*, Cambridge University Press, Cambridge 1988.
46. Jared M. Diamond, *Collapse: How Societies Choose to Fail or Succeed*, Penguin Books, New York 2005.
47. Figure 2.3 from Climate Change 2007: *Synthesis Report*. Contribution of Working Groups I, II, and III to the Fourth Assessment Report of the Intergovernmental Panel on Climate Change (R. K. Pachauri and A. Reisinger (Eds.). IPCC, Geneva, Switzerland, p. 38.
48. International Energy Agency (IEA), *Redrawing the Energy – Climate Map*, World Energy Outlook Special Report, OECD/IEA, 2013.
49. The relative Global Warming Potential (over a period of 100 years) of CO_2, CH_4, and N_2O are, respectively, 1, 21 and 310 (see Reference 18, p. 76 and references cited therein); that is, one CH_4 molecule causes as much damage as about 21 molecules of CO_2, while one molecule of N_2O causes as much damage as about 310 molecules of CO_2.
50. MIT Joint Program on the Science and Policy of Global Change, 2012 Energy and Climate Outlook.
51. Lennart Bengtsson, *The Status of Climate Change Research*, Energy 2050 (p.13), International Symposium on fossil-free energy options, Stockholm University, Stockholm, Sweden, October 19-20, 2009, organized by the Royal Swedish Academy of Sciences.
52. European Academies Science Advisory Council (EASAC), Policy Report 22, *Trends in Extreme Weather Events in Europe: Implications for National and European Union Adaptation Strategies*, November 2013 (ISBN: 978-3-8047-3239-1).
53. The Royal Society and the US National Academy of Sciences, *Climate Change, Evidence & Causes*, Publication, 2014.
54. International Energy Agency (IEA), World Energy Outlook 2008.
55. An inevitable by-product of the advanced activities of modern civilization is heat loss into the environment, that is, thermal pollution.
56. Jianguo Liu and Jared Diamond, Science **319**, 4 January 2008, pp. 37–38.
57. World Economic Forum, *Energy for Economic Growth – Energy Vision Update 2012.*
58. Zhang Guobao, Reference 57, p. 12.
59. Eli Kintisch, Science **318**, 26 October 2007, p. 547; Hao Xin and Richard Stone, Science **327**, 9 March 2010, pp. 1440–1441.

60. International Energy Agency (IEA), World Energy Outlook 2012.
61. Thomas. L. Friedman, *Hot, Flat, and Crowded*, Farrar, Straus and Giroux, New York 2008.
62. Jianguo Liu and Wu Yang, Science **337**, 10 August 2012, pp. 649–650.
63. Abundant supply of natural gas spurs greater use for electricity generation. In the USA by 2035 electricity generation using natural gas may surpass that using coal.
64. Reference 65, pp. 316–320.
65. Niall Ferguson, *Civilization – The West and the Rest*, The Penguin Press, New York 2011.
66. International Energy Agency (IEA), World Energy Outlook 2010.
67. Reference 65, p. 320.
68. Science **339**, 1 March 2013, pp. 1018–1019.
69. http://www.ipcc.ch/publications_and_data/ar4/syr/en/mains1.html.
70. *Ecosystems and Human Well-Being: Synthesis*, Millennium Ecosystem Assessment, Island Press, Washington, DC 2005 (ISBN: 1-59726-040-1).
71. Richard A. Kerr, Science **334**, 25 November 2011, p.1040.
72. See, for instance, IPCC 2007 and 2013; http://www.dnva.no/binfil/download.pdp?tid=58783.
73. Congressman Roscoe Bartlett (www.bartlett.house.gov/EnergyUpdates) August 15, 2007.
74. Robert M. May, *Stability and Complexity in Model Ecosystems*, Princeton University Press, Princeton 2001.
75. Jean-Claude Debeir, Jean-Paul Deléage and Daniel Hémery, *In the Servitude of Power- Energy and Civilization through the Ages* (translated by John Barzman), Zed Books, London1991.
76. *NASA Earth Observatory images by Joshua Steven using Suoni NPP VIIRS data from Miguel Román, NASA's Goddard Space Flight Center.* https://www.nasa.gov/feature/goddard/2017/new-night-lights-maps-open-up-possible-real-time-applications.
77. International Energy Agency (IEA), United Nations Development Programme (UNDP), *Energy Poverty: How to Make Modern Energy Access Universal?* Organization for Economic Cooperation and Development, OECD/IEA, p.12, September 2010.
78. Brian Heap, *Towards Sustainable Production and Consumption,* Proceedings of the Academy of Athens, Vol. 86 A′, Athens 2011, pp. 129–137.
79. Reference 65, p. 210.
80. Λουκάς Γ. Χριστοφόρου, *Βιώσιμη Ενέργεια*, Πρακτικά της Ακαδημίας Αθηνών, Ακαδημία Αθηνών, τ. 83A', 2008, σελ. 3–30; Loucas G. Christophorou, *Sustainable Energy,* Proceedings of the Academy of Athens, Vol. 83 A', 2008, pp. 3–30.
81. Leon R. Glicksman, Physics Today, July 2008, p. 35.
82. G. I. Meijer, Science **318**, 16 April 2010, pp. 318–519.
83. International Data Corporation, Document No. 221346 (2009); www.idc.com.

84. Joint Research Center (JRC), *Science for Water*, JRC Thematic Report, JRC71148, Brussels 2012.
85. The huge future urban influx will require energy efficient ways to purify drinking water. Even when there is water, it takes energy to pump it, to transport it to where it is consumed, and to treat it.
86. Biotechnology could also produce plants for animal feed with modified composition that increase the efficiency of meat production and lower methane emissions.
87. C. Juma, Nature **479**, 24 November 2011, pp. 471–472.
88. http://www.economist.com/node/21602757/print; The Economist, *Digital Disruption on the Farm*, May 24, 2014. Monsanto, the world's largest hybrid-seed producer, has a library of hundreds of thousands of seeds and terabytes of data on their yields.
89. In Europe there is mistrust in the policies of public authorities and firms involved in the commercialization of GMOs and opposition to GM food because it is perceived as an invasion that is uncontrollable, involuntary and a threat to its culture. While in this debate there seems to be a lack of appreciation of the fact that absolute safety is not achievable on virtually anything, risk should be minimized; furthermore, the concern is a function of the nature of the risk involved. Risks directly relating to human health and potential long-range effects are primal.
90. In 2010 "a Eurobarometer survey" (see "Science and Technology", Eurobarometer Special Survey, No. 340, Brussels: European Commission, June 2010, p. 19) indicated that 50% of the Europeans agree with the statement "Europeans feel most strongly that scientists cannot be trusted to tell the truth about controversial scientific and technological issues", while only 16% answered that they disagree with it.
91. Statement of EFSA (European Food Safety Authority), EFSA Journal 2012; 10 (11): 2986.
92. Qiang Wang, Nature **519**, 5 March 2015, p.17.
93. The Royal Society, *Sustainable Biofuels: Prospects and Challenges*, January 2008 (ISBN: 978 0 85403 662 2). See also The Royal Society, *Genetically Modified Plants for Food Use and Human Health – an Update*, Policy document 4/02, February 2002 (ISBN: 0-85403-576-1).
94. Worldwide-agricultural-model-estimates of greenhouse gas emissions from land-use change for biofuel production, has found (see endnote 95) that "corn-based ethanol, instead of producing a 20% savings, nearly doubles greenhouse gas emissions over 30 years", raising questions about large biofuel mandates.
95. T. Searchinger et al., Science **319**, 29 February 2008, pp. 1238–1240.

Chapter 3
Science

3.1 Introduction

Science today studies phenomena lasting less than 10^{-21} s and phenomena which occurred more than 13 billion years ago; science also studies phenomena occurring over distances greater than 10^{28} cm and shorter than 10^{-13} cm; that is, science studies phenomena occurring over times and distances varying by a factor of about 10^{40} (Fig. 3.1).[1] In those incomprehensible ranges of time and space, the description of the physical world presented by science is most impressive. Based on what we know today everything in this enormous cosmos everywhere is made up of the same microscopic particles, the atoms and their constituents; their behavior is governed by the same physical laws everywhere.

Progressively science presents to us concepts and objects of the physical world which are beyond our usual experience. In fact, most of modern science lies beyond our vision and senses. In science, we learn to know without seeing and what we indirectly see is wonderful in spite our indirect contact with it. For 50 years,[2,3,4,5] I have been studying the slow electron either bound in atoms and molecules or moving freely in gases, plasmas, electrical discharges, accelerators, scientific instruments or quasi-freely in condensed matter; I released it from surfaces, ejected it from atoms, molecules and ions, attached it to atoms and molecules, detached it from negative ions, had it collide with molecules, ions, atoms, and used it to study new features of the atomic and molecular structure – yet I had never seen it. I saw it and I understood its behavior only through its specific properties that identify it as a particle (with specific mass, charge, spin, energy, de Broglie wave length, etc.), and this is scientifically sufficient, for whether modern science presents to us a picture of the physical world, or a picture of our relation to the physical world, our indirect contact with the physical world as it is provided by science is equally beautiful and meaningful.

L. G. Christophorou, *Emerging Dynamics: Science, Energy, Society and Values*,
https://doi.org/10.1007/978-3-319-90713-0_3

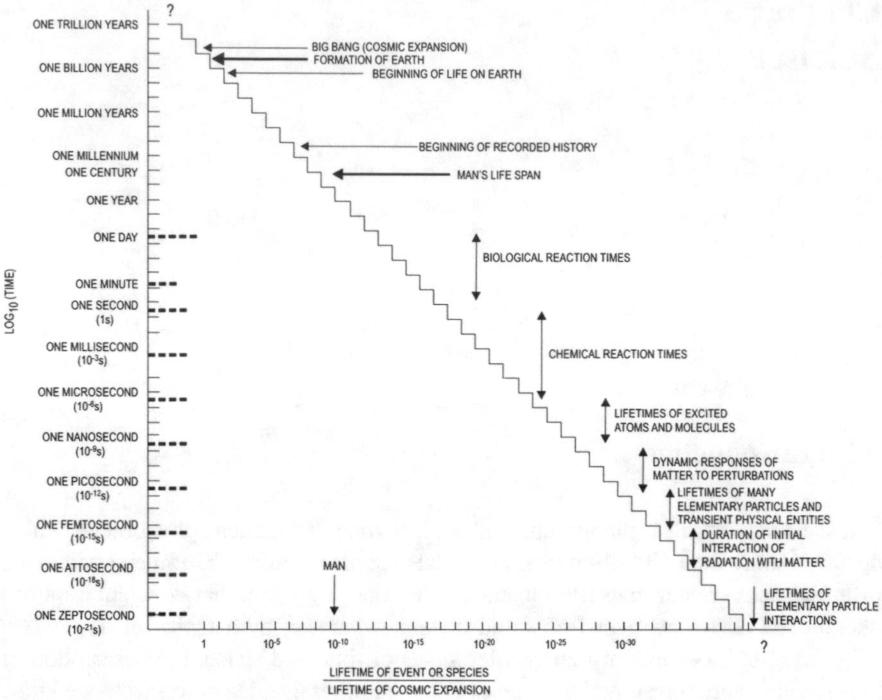

Fig. 3.1 Lifetime or duration of physical/chemical/biological interactions or phenomena as a function of the ratio of the lifetime or duration of the phenomenon and the age of the universe since the big bang. The scales are logarithmic (every division corresponds to an increase in time by a factor of 10) (based on Fig. 1.3 in Reference 1)

3.2 Meaning of Knowing

Verification of knowing must have a source which itself is not in knowing. We need to know what we know and we need to know how well we know what we know, and we need to allow for the things we do not know that might exist. Knowledge can only be looked at from the point of view of the method of knowing. Whether we know through deduction, reduction-induction, or holism, or through inference and faith we are limited. This then instructs caution for all knowledge carries its own limits and is subject to doubt.

3.2.1 *The Inductive Method of Science*

Our knowledge about the physical world began with the sixth century BC when Thales of Miletus (c. 620–c. 546 BC) asked "What is the world made of?" and he answered, "Water". The question is still being asked and the answer today is of course different, "Energy" (e.g., see References[6,7], Chaps. 1 and 7 and Appendix).

The "natural" philosophers of Ionia – among them Thales of Miletus, Anaximander (c. 610–c. 546 BC), Anaximenes (c. 585–c. 528 BC), Heraclitus (c. 535–c. 475 BC) and Anaxagoras (c. 510–c. 428 BC)[8] – were the first to look for simple principles behind the variety of the observed physical phenomena. The principles sought then were substances; today they are theories and "laws".

The early natural Greek philosophers dealt with generalizations. Heraclitus taught that there is an underlying unity of the world and argued that the fundamental essence of the cosmos is *change*; he emphasized the generality and continuity of change. "Τα πάντα ρεί" (everything is in a continuous state of flux, in a perpetual motion like a river) he said. Like the old Greek philosophers, the modern physicist reaffirms: nothing is at rest; science deals with change; we live in a restless world permeated by ceaseless change everywhere.[9]

A little later, and in opposition to Heraclitus, the philosopher Parmenides of Elea (c. 515–c. 460 BC) argued that the fundamental structure of the cosmos is unchangeable; *the being* ***is*** and it exists as a whole.

The ancient Greeks accommodated Heraclitus' and Parmenides' diametrically opposing views in the most brilliant way, that of *complementarity*. The world *is*, but it *is continuously changing, perennially becoming*.

The holistic approach of Heraclitus and Parmenides was followed by the opposing view of the atomists, foremost Leucippus (c. 480–c. 400 BC) and Democritus (c. 460–c. 370 BC), who taught that the world is made up of definite unchangeable and indivisible substances – which they called *atoms* – moving in a void; the combination and separation of these atoms leads to the visible matter and causes the perennial change and transformation of the cosmos. Democritus' atomism was the first example of reductionism: *from the cosmos to the parts – the atoms – and from the parts – the atoms – to a construct for the cosmos.*

Atomism itself was opposed by Aristotle (384–322 BC) because it was considered not suited to describe, let alone explain, the order, harmony and the purpose of the cosmos. Aristotle's philosophy was centered on holism and teleology. The whole he concluded deductively is more than the sum of its parts and it has characteristics of its own which are not reducible to the properties of its constituent parts. There, thus, has been present in ancient Greece a deep conflict between holism (the understanding of the whole as a unified entity) and reductionism (the understanding of the whole based on the knowledge of its parts), which persists to this day.

Greek philosophers, principally Plato (427–347 BC), taught that truth adheres to *axioms*, a priori concepts in the "world of ideas" with indisputable validity. Plato believed that the whole structure of the cosmos could be deduced from such axioms and developed the concept of *deductive reasoning* and the logical techniques for proof. In some way, this approach is analogous to the thinking of some theoretical physicists today who believe that truth is to be found in logical mathematical structures, in mathematical models.

Deduction has since become the philosopher's powerful tool to the truth, its basic limitation being the universal validity and self-evidence of the axiom. Deduction makes explicit information which already exists in the axiom; the validity of the deduced entity is derived from that of the axiom.[10,11]

The method of modern science is both deductive and inductive. While deduction is a powerful tool to the truth, the power of modern science comes from its inductive method based on the ability of science to reduce and the use of the experimental method to "interrogate" nature under reduced and controlled conditions. This way observations and experimental data are checked for consistency and are harmonized among themselves; they are validated by the known physical laws and are explained and unified by general concepts and theories: *From the experiment, inductively to a general theoretical rationalization, and from the theory deductively to the results of the experiment and to more experiments.* In this two-way inductive-deductive process the agreement between theory and experiment and the experimental confirmation of the predictions of the theory constitute the fundamental criterion for the validity of the theory and its conceptual formalism and foundation.

Systematically, modern science has reduced the physical world to molecules, atoms, nuclei and subatomic particles and based on this knowledge at the most fundamental reduced level succeeded in establishing the physical law and the order of the physical world, based on the validity of the physical law. Through this "reductionist" approach, physical science laid the ground for a unified description of the physical world at the microscopic and, by extension, at the macroscopic scale.

The recognition of the power and the limits of this extension – that is, the recognition of the fundamental interconnections between the properties of the nuclei, the atoms, and the molecules on the one hand and the various forms of matter, the world and life, on the other hand – is most significant. This is because the reduction of the physical world to its elemental constituents and the discovery of the fundamental laws that describe their behavior at the extreme reduced and isolated level of matter, does not lead to its reconstruction based on ***only*** this knowledge. It is hindered, for instance, by the enormous differences in scale and complexity and the interactions a system continuously undergoes with its environment. This is especially significant when one refers to living organisms the behavior of which may be rationalized by new laws over and above the known laws of physics which describe the behavior of their constituent parts. The whole is manifested in the particular, but it has properties of its own.[12,13]

As in ancient Greece, many scientists today have stressed the need to consider holistic properties and teleology issues and have pointed out that neither it is possible to comprehend the basic phenomena of life based on the known laws of physics, nor it is possible to understand man only by a reductionist view of him. The preponderance of scientists today, however, hold the reductionist view, that "*all the explanatory arrows point downward*",[14] that is, everything can be boiled down to molecules, atoms, and particles, life itself included. *This reductionist view of man conflicts with the world-view of a large fraction of society.*

The history of science teaches (see endnote 1) that science progresses in small steps normally based on observations and experimental data and their correlation via the inductive-deductive method – small steps which demonstrate the validity or the inadequacy of existing theories; in the latter case, they call for the introduction of new concepts and new theories. Small steps like the observations of the astronomers of ancient Greece, the calculations of Kepler, and the experiments of Galileo that led to the laws of classical physics of Newton. Small steps, also, like the first experiments which determined the nature of electricity and magnetism, the discoveries of Faraday, the laws of Coulomb, Ampere and Gauss which themselves led to the synthesis of Maxwell, the existence of electromagnetic waves and the speed of light. In a similar fashion, small steps like the absorption and emission spectra of atoms, the ionization of gases by ultra violet radiation, the light emission from hot objects, the discovery of radioactivity, and the discovery of the electron, which led to the special theory of relativity, the quantum theory of light and matter, and the quantum mechanical description of the microcosmos at the most fundamental level. An important part of this process has been the introduction of new concepts such as the concept of the field, electromagnetic wave, photon, quantization of energy, relativity and so on.

Similar schemes can be sketched for biology. One can identify, for instance, the following four important steps in the evolution of modern biological sciences (see endnote 1)[15]: The theory of evolution in 1859 by C. Darwin and A. R. Wallace; the discovery that *DNA* is the genetic material in 1944 by O. T. Avery, C. M. Macleod and M. McCarthy; the discovery of the molecular structure of *DNA* in 1953 by J. Watson and F. Crick and others; the sequencing of the human genome at the very end of the twentieth century by groups of scientists. In biology, the nature of the scientific concepts employed is radically different from that in physics: *evolution is a macroscopic concept and DNA is an incredibly complicated macromolecular structure.*

3.2.2 Reductionism and Holism

Our understanding of the physical world in its extreme simplicity as is provided by the reductionist method of science far exceeds our understanding of Nature at higher levels of complexity. This is because as the complexity of matter increases gradually emerge new properties of matter and new physical phenomena which cannot be accounted for or be predicted by the reductionist method of science. That this is so can be illustrated by the following examples.

From isolated particles to a system of particles At the subatomic level, matter consists of a variety of types of particles. While the particles of each type are similar and interchangeable, their behavior as a system cannot be analyzed based on their individual behavior alone. In fact, the behavior of a system of particles depends on the kind of particles the system consists of. Let us consider the kind of particles called "fermions" and specifically the electrons which have a spin equal to 1/2. A collection of electrons obeys Fermi statistics and thus behaves according to the Pauli Exclusion Principle: No two electrons with the same values of their four quantum numbers can occupy simultaneously the same quantum state of a polyelectronic atom. The principle describes the fact that electrons in atoms cannot spiral down into the nucleus, but instead they form a series of shells which surround the nucleus, giving atoms their distinct structure. The number of electrons in the outermost atomic shell produces the chemical valence, which in turn binds atoms into molecules. The properties of the system of these outermost electrons define the kind of atoms and molecules that can exist and thus the kind of materials found in nature.[16] The Pauli Exclusion Principle applies to systems of particles and has little meaning for individual particles. The features it describes are not mere extensions of the properties of the individual particles, although they depend on them (see endnote 16).

Let us look at another type of particles where matter consists of a very large number of neutral molecules, as in room-temperature air (where at standard pressure and temperature the number density of molecules is $\sim 2.7 \times 10^{19}$ molecules cm^{-3}). The thermodynamic behavior of this system of particles obeys Maxwell-Boltzmann statistics and the concepts of gas temperature and gas pressure define experimentally measurable macroscopic properties of the gas (the air). While the macroscopic properties temperature and pressure of the gas are related to the microscopic motions of the individual molecules making up the gas, they are meaningless for the isolated gas molecules; they are new, *emergent properties* of the whole collection of molecules making up the gas. The temperature and the pressure are properties of the entire system of particles and not properties of the individual molecules themselves.

From isolated atoms to isolated molecules Atoms form chemical bonds between them when they come sufficiently close together because the *emergent composite structures*, the molecules, are energetically more stable than the separate atoms comprising them. The atoms make something new and different – chemical compounds – with new properties which depend on the properties of the atoms making up the compounds but are not reducible to them. Molecular structure is more complex than the atomic. For instance, the hydrogen molecule (H_2) is formed when two hydrogen atoms come close enough together that the orbits of their electrons overlap and the probability of the two electrons being in the same space between the two hydrogen atoms of the hydrogen molecule is large. For these conditions to be satisfied, the electron of the one H atom must be moving in the opposite direction than the electron of the other H atom, that is, the spins of the electrons of the two H atoms must be anti-parallel. Only then the probability of finding simultaneously the two electrons between the two H atoms is large and the negative charge of the two electrons between the two nuclei attracts them together, and a bond between the two H atoms is established. In the opposite case, where the spins of the two electrons of the two H atoms are parallel, the electrons cannot be simultaneously between the two nuclei, the force between the nuclei is repulsive, and the H_2 molecule is not formed. This is what is demanded by the law of the lower level of complexity for the electrons.

In the hydrogen molecule, the two electrons not only have different energies, but the hydrogen molecule has different absorption and emission spectra, different ionization energy and new characteristics such as vibrational and rotational structure. The laws of the lower-level complexity systems are seen to determine the kind of higher-level complexity systems possible, but the properties of the higher-complexity systems have no analogy to those of their parts.

Let us consider another molecule of still higher-level of complexity, that of water (H_2O). The chemical and physical behavior of the water molecule is determined by its quantum mechanical structure, which itself is dependent on the electronic structure of its constituent atoms H and O. The water molecule has physical and chemical properties of its own, for instance, an electric dipole moment. In turn, the electric dipole moment and the stereo-chemical structure of the water molecule crucially determine its interactions with other molecules. Water molecules make "hydrogen bonds" with other water molecules (hydrogen-bonded structures), transient clusters comprised of specific numbers of water molecules, complexes around positive ions, around negative ions or around electrons (forming "hydrated" electrons), and so on.[17]

If a few trillion trillion water molecules are put in a glass, the whole assembly of the water molecules, the liquid, acquires a new property, *fluidity,* that none of the water molecules has. Fluidity, wrote Phillip Anderson in an article with the title "More is Different", is an *emergent* property of liquid water.[18] If, now, liquid water is heated up to 100°C, the same molecules evaporate and the system makes a phase transition to water vapor. If, conversely, liquid water is cooled down to 0°C, the system abruptly undergoes another phase transition, the water molecules stop their chaotic motion and form an ordered hexagonal crystal structure known as ice. These forms of matter are *emergent*; they have no meaning for the isolated water molecules. The transition from the liquid state of matter to the crystalline forms of matter is normally associated with two important *emergent* properties: *order* and *organization.*[19]

Many other similar examples can of course be given and indeed one can extend this type of discussion even to our Galaxy and the entire universe. The whole, the Milky Way Galaxy, with its unparalleled complexity and dimensions has new. *Emergent*, properties – for example, black holes – above and beyond those of its constituent parts; incredible plethora of new phenomena and new properties arising from Nature at this scale of size and level of complexity.

From inorganic matter to the phenomena of life The reductionist quest for knowledge in science discussed in the previous sections, is faced with fundamental problems of principle in accounting for the behavior of biological systems. As we have repeatedly said, reductionism deals with the parts, holism treats the systems as wholes. We cannot comprehend the whole without the knowledge of the parts, but we can comprehend the parts without comprehending the whole. In biology, the goal is to understand not what things are made of, but how they are put together and function as integrated wholes; how totally new structures (e.g., the embryo) emerge in the progression from inanimate to animate matter (see endnote 19).

The view that life has been reduced to its molecular basis is thought to have been strengthened by the great recent advances in biochemistry, molecular biology and genomic science. Large molecular structures are normally "loose" systems with many weak bonds between their atoms and with many degrees of freedom which facilitate their transformation into other structures through a multiplicity of rearrangements, mechanisms and interactions that allow their co-evolution with their surroundings. At every level of molecular complexity emerge new structures which change the ability of the system to evolve. In this way, knowledge of the structure of biological molecules becomes the basis for understanding their biological action. It is known for instance that the biological action of macromolecules is affected by the "structured" water of the cell milieu[20]; for example, it affects the way in which macromolecular structures fold and change their shape and, consequently, their reaction mechanisms.

The road from physics to biology is thus obstructed by the extreme complexity of the systems of biology. Many distinguished scientists (e.g., Phillip Anderson (see endnote 18), Edward Wilson,[21] Paul Davis (see endnote 19),[22] Eugene Wigner,[23] Stuart Kauffman[24,25]) support the view that the understanding of the holistic properties and the organizational principles of biological systems will likely be achieved with autonomous laws which deal with complexity and auto-organization of matter, laws above and beyond the known laws of physics. Such basic laws remain unknown.

There remain countless fundamental questions in going from the physical and biological understanding to that of the human person.

3.2.3 The Indirect and the Complicated

The questions about the elements the world is made of, the nature of being and becoming, reductionism and holism, facts versus constructs and ideal forms, creation of constructs, inductive and deductive reasoning, and the way we go from the particular part to the whole, from the simple to the complex, from the small to the big, from the short-lived to the long-lived, from the microscopic event to the macroscopic effect, are all still open-ended; more often than not they are partially answered or unanswered.

Even the experimental measurements we so much rely upon are often convoluted with many variables that make "the experimental results" indirectly-deduced quantities, often remotely connected to the initial measurement or event. It used to be that a scientist's experimental equipment was simple enough for him to claim that it was an extension of himself in his pursue of knowledge. There used to be a time in the past when the scientists made their measurements, analyzed them, interpreted them, deduced their conclusions and little else; all was personal and direct. Today, with very few exceptions, this is so no more: all is impersonal, indirect and remote. Both modern experiments and theoretical computations and simulations are too complicated to keep direct control of, and the knowledge acquired too remote and too indirect. And although, and despite its abstraction and indirectness, such knowledge is "real", it remains a fact that *the new methods of science make direct contact with nature difficult; new methods of learning tend to provide convoluted information.*

Look, for instance, at supercomputing (soon on its way to the exascale, computers that can perform a billion billion (10^{18}) calculations per second). Supercomputing is becoming an essential tool not only in physics, chemistry, fluid mechanics, astrophysics, cosmology and material science, but in neuroscience, biological and life sciences, climate change simulations and predictions, and so on. The convoluted nature of the knowledge it provides, often makes direct experimental confirmation difficult. Information technology is revolutionizing how research is done and how researchers interact with each other.

3.3 The Nature of Truth and the Image of Reality

Every age of history seeks "the truth". But what is truth? What is the essence of truth? To this old question there is still no simple answer because the answer depends on the context within which the question is being asked and on the method of knowing being used to know. If, for instance, we accept that truth is the concern of both science and faith, what is the nature of the two kinds of truth and what is the relation between them? Independently of the answer, truth is related to freedom and freedom is the essence of human existence. We are, then, instructed to be tolerant of and receptive to the diverse ways in which truth is mirrored in all things and to the methods of knowing.

Science is underpinned by the belief that there is a truth about the physical world that Nature can be made to yield, if only one knows the proper questions to ask. Scientific results then are regarded as something "found by" or "disclosed to" the scientist. Van Fraassen[26,27] sees scientific realism as a belief in a "deep structure of reality", to be revealed by scientific inquiry. We must, thus, in principle, be able to recognize the truth when we reach it, if not before. There is, however, also, the view that "scientific results are to be construed as imprints made by the human mind upon Nature", a view that accepts a level of relativism and subjectivism in the knowledge of scientific truth. Such a view is expressed in the writings of Polanyi[28] and Kuhn[29] and is apparent in the conjectures by Einstein and Bohr about aspects of scientific reality.[30]

Truth is central to science: you can practice science only if you respect the truth. But who defines scientific truth? Scientific truth could be defined as that which scientists affirm and believe to be consistent with the accepted body of scientific knowledge. Scientific truth, then, is found, not given, and it is tested, "subjected to verification". But not all truth is verifiable or complete. Even Aristotle's argument that truth is the agreement with the facts of what is being asserted is not always possible to ascertain. And there are still those[31] who wonder if in a changing universe, the physical laws remain unchanging. Thus, ultimate, absolute, complete true knowledge is not in the court of science; even in science truth is elusive and the true nature of reality may not be amenable to knowing. Although it is in the nature of charge to be subject to the Coulomb force, writes Cartwright,[32] this nature does not in any way reveal *the essence* of charge. Yet, there is scientific knowledge (truth) so well-defined and scientific explanations so thoroughly tested and "confirmed" that they are confidently held and are indeed transformed into powerful useful science-based technologies.

In man's search for truth, the problem lies not with the scientific findings themselves, but rather with the view that there is no truth other than that provided by science. But if a thing cannot be subjected to scientific testing is it necessarily wrong or unimportant? Should we insist on one truth or should we rather concede that there are many kinds of truth, or various aspects of truth, accessible via complementary ways and methods of knowing? Truth we believe is more than what can be accessed by any single known method of knowing. Beyond that which we know with certainty, lay the vast ocean of the unknown and the unknowable, and all knowledge is shrouded in doubt. We are then led to conclude that the wholeness and the unity of truth presuppose true complementarity among the various kinds of knowledge and the methods of knowing.

3.4 The Laws and Concepts of Science (Physics)

Let us look at the laws of physics and the concepts and constructs behind them. The laws of Nature, as established by physics, are mostly inductive generalizations of reductionist knowledge; precise quantitative relationships between physical entities found by repeated experiments and observations, reflecting persistent regularities in the behavior of the physical world, which are rationalized deductively with reference to a broader law based on a concept, usually mathematically structured. Concepts, maintains Barrow,[33] are more profound than the physical laws; in the concept of gravity, he writes, "we express not a specific law of physical behavior, but a unitary picture of how Nature works: what holds the world together and yet allows it to move and evolve."

Behind every law of physics there is a concept and, thus, every interpretation and rationalization of physical facts and phenomena, and, consequently, every comprehension of Nature based on the physical law, depends on human concepts. Reality, however, does not owe its existence to concepts. No concept is final; concepts are made and remade, and new concepts evolve as new knowledge is acquired. Such, for instance, is the sequence of concepts from classical physics (particles, waves, forces, etc.), to gravity, electromagnetism, quantum physics, relativity, string theory. True revolutions in science are transformations of the concepts upon which science is based (such has been the case in the work of Galileo, Newton, Maxwell, Planck, Bohr, Einstein and so on) and the resultant unifying power of science. Since all human constructs are subject to change, all knowledge based on physical law however elegant is temporal and incomplete. The physical laws are not rigid; time and again, one physical law leads to and is superseded by another more general and more precise. Nonetheless, the established physical law represents enormous compression of information and on it rests the power of science to predict.

The laws of physics apply across all of science and appear to hold in every part of the universe so far investigated, and in that sense, they are presumed to be universal. They, however, have been established by looking at the physical universe 13.8 billion years after the big bang and are based on knowledge of inanimate matter.

Source (Object) →Field →Force →Energy Transformation and Flow

Fig. 3.2 Fields relate to sources and, through forces, to energy transformation and energy flow

The universality of the physical law – holding for everything, everywhere, and for all times – is, thus, implicit. The physical laws we now know, for instance, may not be applicable under the extreme conditions at the very beginning of the universe. It should perhaps be noted also that the laws of macroscopic matter are seemingly unaffected by the laws of microscopic matter, and, conversely, that the laws of microscopic matter are seemingly unaffected by the laws of macroscopic matter.

Let us then look, by way of example, at just the concepts of the *field* and the *force* which are relevant to the concept of *energy* so frequently referred to in this book. The field relates to a source (an object) and, through the force, it relates to energy transformation and energy flow (Fig. 3.2).

Sources of fields are charges (positive and negative) and masses. Charges or masses entering the field created by the charge or the mass experience a force and gain energy. Fields do not occupy space, but extend throughout space, including vacuum; they contain energy and their strength diminishes as the distance from the source increases.[34] Fields are invisible; they are "seen" by their effects.

Stationary charges generate electrostatic fields and *moving* charges (electric currents) generate magnetic fields, and if a magnetic field changes, it generates an electric field. Electric and magnetic fields can be coupled, constituting two parts of a greater whole – an electromagnetic field. The classical concept of an electromagnetic field is one of a smooth and continuous field, which extends indefinitely through space propagating in a wave-like manner, exerting a force on other charges via the so-called electromagnetic interaction. The electromagnetic wave has energy, which is proportional to the frequency of the wave. This classical concept of the electromagnetic field is complemented by the quantum-mechanical concept of the electromagnetic field as a quantized entity, comprised of individual particles, quanta or photons, each having a fixed energy $E = h\nu$, where h is the Planck constant and ν is the photon frequency; in this picture, then, energy moves discontinuously.

Similarly, mass is a source of field. A mass object establishes a gravitational field around it. The gravitational field created by all the mass in the universe, gravity, pulls on every particle of matter in the universe. Just as in electromagnetism moving charges generate electromagnetic waves, in the theory of general relativity moving masses generate gravitational waves. However, because of the weakness of gravity,[35] astronomical amounts of matter must be moved around to generate waves on a scale that might be detected. Such gravitational waves from a merging two black-hole system have recently been detected.[36,37]

Fields give rise to forces and forces play a vital role in understanding the interactions between the constituents of matter and energy. Indeed, today, all interactions in nature are studied in terms of four fundamental forces – gravity, electromagnetic, strong nuclear, and weak nuclear – mediated by the exchange of particles. What does force have to do with energy? *Forces cause the transformation and flow of energy.*

An increasing number of scientists see the reductionist laws of physics to be limited when applied to the structures of biology and as inadequate to explain the behavior of the complex systems and functions of living organisms (see endnotes 23–25). Wigner (see endnote 23) maintains that the laws of physics would have to be modified drastically if they are to account for the phenomena of life, and Kauffman (see endnotes 24, 25) advises "to go beyond reductionism into emergence."

An increasing number of scientists are then led to ascertain the possibility that there are laws which govern the behavior of complexity and living matter which need well-developed concepts and constructs and a new vocabulary that might include terms like emergence, organization, information, growth, adaptation, genes and so on, suited for biology rather than for physics. These laws are envisioned to be fundamental in the sense that they cannot logically be derived from the underlying laws of physics.

3.5 Distinct Characteristics of Science

Science is the most successful method that has ever been employed for the understanding of the physical world; knowledge obtained by science can in most cases be tested and can thus be validated. There is only one science and the knowledge it provides is society's heritage of common knowledge. As such it transcends national boundaries and generations. It belongs to all humanity; generation upon generation builds on all of humanity's prior scientific accomplishments. Although today most of the scientific research is still done, and most of the profit from the latest advances in science is still concentrated, in the developed nations of North America, Europe, and Asia, scientific knowledge is, in principle, available to everyone; modern communication systems have made this possible. The heritage of common knowledge provided by science is a unifying force for humanity.

Embedded in the tradition and method of science are distinct characteristics of science which qualify many of its functions. These characteristics need to be recognized and adhered to if science is to serve well and in a balanced way both itself and society.

Science is a self-correcting system Science is cooperative and at the same time encourages originality, independence and dissent. It stresses the need for an open mind; time and again the scientist must reverse direction, and he normally does. Proven scientific positions proved wrong no matter how great. Interpretations of experimental data and observations, explanations of events, and paradigms of theories have had alternative rationalizations and have always been limited, never complete. This helps the scientist tolerate ambiguity, strive for improvement, and allow for error and self-correction.

Science teaches the value of relatedness and embeddedness A necessary condition of all life is interdependence; everything relates to everything else; nothing exists in isolation. Hence everything assumes essence via its interactions with the something else. Science, therefore, seeks not only truth, but also relatedness and embeddedness within its domain. One branch of science relates to and in varied degrees is embedded in another. Out of this implicit coupling of the parts of science emerges the underlined unity among its seemingly chaotic functions. The mutual embeddedness of the parts of science allows for their integration, feedback, and accommodation (e.g., embeddedness of physics in chemistry, biology and technology and vice versa). Each branch of science, especially the neighboring ones, cross-fertilizes the others; they draw from, reinforce, and are indebted to each other. This process is continuous and accounts for the unceasing readjustment of the functions of science within and between its parts and the ultimate cohesiveness and advancement of science as a unified whole.

Science is changing, it is becoming more complex The growth of modern science is becoming increasingly more complex. How will this affect its future efficiency, stability and resilience? Will science get so complex that it will collapse like other things do when they have no other way but to keep growing and to keep becoming more complex, or will science advance independently of its size and complexity? Clearly, this would depend on its practitioners and the users of scientific knowledge, but it would also depend on the big politics of science, science's big sponsors, and on how society perceives science and its impact. Large international projects like the Intergovernmental Panel on Climate Change, the human genome, the International Thermonuclear Experimental Reactor, the European Organization for Nuclear Research, the large "user" facilities (e.g., particle accelerators), the big data facilities for medicine, and the knowledge being generated in cyber space, are but examples of this trend. Even the character and culture of today's large-scale research at major research facilities has been changing.[38]

Changes in the production of scientific knowledge, including the growth of "hybrid" public-private sponsorships raise concerns about the independence and impartiality of science. Similar-type concerns are raised by changes in the way science and science-based technology are governed and operated, particularly because of globalization and the sheer size of some of these activities. The earliest and possibly strongest concern has been the "Military-Industrial-University Complex". It has since been joined by other "complexes" such as "The Medical-Industrial Complex",[39] which is seen to be increasingly shifting research done by universities to companies. The shifting of activity in biomedical science toward the big projects and the big companies continues unabated.[40] For instance, genomics is driving megamergers as companies seek to lock in patents and licensing agreements, and so do agrocompanies. Many worry that this shift will further erode science's independence and impartiality.[41]

Clearly, a fundamental change is seen in the functioning of science showing that the future of science rests not only with its practitioners but perhaps more so with its sponsors and its users.

A most distinct characteristic of science is its universality which is discussed in the next Section.

3.6 The Universality of Science

Science is universal in at least two fundamental ways: First, regarding the applicability and validity of its method, the generality of the physical law, and the effects of scientific knowledge on human functions and wellbeing. Second, regarding the participation of humankind in it; science's growth is rooted in the discoveries of all nations, and the knowledge it provides is (or can be) universally-shared. A prerequisite for the universality of science is freedom of work and communication in science, wisdom and caution in the application of scientific knowledge, and opportunity for every nation and every generation to participate in, and profit from, scientific discovery. This way, the heritage of common knowledge provided by science becomes a unifying force of humanity and a source of universal hope. Meaningful participation in a broader effort and sharing in the common accomplishments instills pride in the individual *whether* in science or in other walks of life. Through the knowledge science provides, science-based technology brakes up the isolation of totalitarian states, liberates oppressed peoples, and exposes human suffering. Indeed, science has made the panoply of totalitarian and dictatorial regimes obsolete by enabling its penetration from within and from outside. Science is, thus, looked upon as a liberating force for humankind (see endnote 1).

The universality of science has been defined by several scientific organizations. For instance, for the International Union of Pure and Applied Physics (IUPAP) the universality of science "entails the free circulation of scientists, the freedom to communicate among scientists and to disseminate scientific information", and for the International Council for Science (ICSU) it entails the "freedom of association, expression, information, communication and movement in connection with international scientific activities". These scientific freedoms should be coupled to responsibilities.

The aforementioned definitions do not seem to be broad enough. As I have indicated earlier (see endnote 13), the definition of the universality of science needs to be broadened to include the *universal acceptance of science by society*; in those lectures (see endnote 13), I have outlined some of the limits to and some of the needs for the universality of science. I had expressed the view that while the recent advancement of science has been spectacular and the scientific frontier endless, the universal acceptance of science by society is still limited and is still in need of a more effective transmission of the intellectual and cultural value of science. A large fraction of society looks at science with fear and suspicion and views a science-dominated world as unbalanced. A good fraction of society also fears science's impact on man and believes that science has "set its own conditions and imposes its own values" on society.[42] In Vaclav Havel's view, "modern science describes a single dimension of reality.... and the fewer answers the era of rational knowledge provides to the basic questions of human being, the more deeply it would seem that people, behind its back, cling to the ancient certainties of their tribe".[43] A view not unlike the harsher one of Jacques Maritain, claiming that the "deadly disease" science set off in society is "the denial of eternal truth and absolute values".[44] Such views restrict science.

3.6.1 Limits to the Universality of Science

I have elaborated earlier (see endnote 13) on the limits to the universality of science and identified six areas which limit science's universality. In this Section reference is made to these limits, which are further expanded.

First Limit: The preponderance of humanity is still not participating in the advancement of science and does not share the fruits of scientific knowledge.

Although today science transcends locality, it still bears the imprint of locality; and although we scientists demand freedom, we often forgo responsibility.

Most of the scientific research today is done by the developed nations of North America, Europe, and Asia. About 95% of the world's Research and Development (R&D) is conducted by the 20% of the technically advanced peoples. Despite recent progress, neither the scientific knowledge nor the scientific technology is available to most of the world. Most developing countries are practically with little or no science. According to a recent report by the InterAcademy Council (IAC)[45] most industrialized countries are devoting between 1.5% and 3.5% of their Gross Domestic Product to R&D and in fact many have pledged to increase these investments.[46] This great expansion of R&D has altered the global distribution of science and engineering. According to a US National Science Board report,[47] world-wide R&D expenditure rose from $522 billion in 1996 to $1.3 trillion in 2009; interestingly, also, while in 1999, 38% of the world's R&D was performed in the USA, 27% in Europe and 24% in Asia, in 2009 Asia accounted for 32% of world-wide R&D, the USA 31%, and Europe 23%.

In today's world, the scientific and technological isolation of most of humanity is not acceptable. Without proper access to scientific literature and technical information, and without adequate means and materials needed for their indigenous science and technology, *developing nations will continue to remind us of the limits of the universality of science.*

The universality of science requires willingness to find the means and to devise procedures that will allow sharing of scientific knowledge. Indeed, how can we call for a *"knowledge society"* and still restrict direct access to scientific literature for a large part of humanity? ***ALL scientific publications should be made public at the instant of publication.*** Scientific literature should be regarded humanity's property, and access to it should be unconditionally free of charge. It is encouraging that recently it is generally acknowledged that the denial of access to a substantial part of the scientific literature without subscription constitutes a serious impediment to the advancement of science. Progress is thus being made.

Second limit: The "limitless power" of the method of science.

While the ability of science to answer questions which can be defined scientifically is practically limitless, *the impression that there is no limit to what science can do, limits the universality of science,* because there are limits even within the borders of science. As it has been pointed out repeatedly in this book, the reductionist approach of physical science cannot explain the properties of living matter; it is hindered by the enormous differences in the scale of complexity and by holistic properties and teleology issues. Live organisms may have their own laws, which, while not in opposition to the known laws of physics, cannot be reduced to them, because the understanding of living organisms is not possible with only the knowledge of the atoms and the molecules that constitute them. Neither it is possible to comprehend the basic phenomena of life based on the known laws of physics, nor it is possible to understand man only by a reductionist's view of him.

There are still other limits at the boundaries of science. There are, for instance, questions which although defined scientifically, have no scientific answers when formulated because they lie outside the province of science (see Chap. 6); questions such as "What is the origin of the universe?"; "What is the origin of life?"; "How did the first organism emerge from inorganic matter?"; "If matter evolved according to the laws and the forces of Nature, what is the origin of those laws and those forces?". These are questions that science can ask, but science cannot answer (at least for now). Scientists can express opinions about such questions, but they cannot provide scientific answers. Questions of this kind show that although the borders of scientific knowledge are continuously expanding, some questions remain; they belong to the area of "trans-science" (see endnote 1).[48] *The pretension that we have scientific answers for such questions, limits the trustworthiness of science and its universality.*

Third limit: The real or perceived adverse impact of science on traditional values; fears that the scientific view diminishes man.

Values lay deep in humanity's multiple cultures, traditions and religions. Traditional values – such as respect for life, liberty and justice; commitment to peace, freedom, and human dignity; reciprocity – are mutually embedded and mutually indebted; they guide human behaviour and constitute the frames of reference for value judgment (see Chap. 5).

Science per se does not deal with values. Science, however, is not value-free in the execution of scientific research and in the application of scientific knowledge. There are values ***in*** science and there are values ***of*** science. The search for truth in science imposes on the researcher a moral conduct, which is not unlike the moral conduct of a person in the broader society, but it goes further. Science confronts the work of a scientist with the work of his colleagues and cannot survive without justice, honour and respect among them. Science, furthermore, is based on the free communication among scientists and on mutual trust. Freedom of thought and speech, justice, dignity, self-respect, and tolerance of differing views are all values recognized in the past – long before modern science – as necessary for the survival and wellbeing of society. Science relies on these very values for its functioning because scientific research is conducted by and for people; because science itself is first and foremost a human activity. Thus, while the scientific picture of the world changes continuously, the values on which science and scientific behaviour are based remain fundamentally the same: universal, timeless values.

There are as well, the values **of** science, which characterize its functioning: rationality, verification of knowledge, discovery and correction of error, respect for and acceptance of the proven fact, unification and coherence of scientific knowledge, cooperation, humanism. Humanism is a uniquely multidimensional value of science; as I wrote elsewhere (see endnote 1), "If deep in the essence of civilization lies the emancipation of humanity, society cannot be truly civilized without science". These values of science need to be broadly appreciated and to be recognized as complementary to the traditional human values. This recognition and this complementarity are necessary to moderate the image of modern scientist as antagonistic to accepted beliefs, norms, and values, and as increasingly questioning the traditional foundation of Western Civilization. *The degree to which the scientists and society are successful in this endeavour will enhance or limit the broader acceptance of science by and for the benefit of society, and thus the universality of science.*

The scientist, furthermore, is faced with the deep-rooted fear of society that the scientific view of life diminishes man. To the Greek philosopher Protagoras "*Man is the measure of all things*" and to Aristotle "*Man is the ultimate supreme creation in the cosmos*", while to Christianity "*Man is the image of God.*" Today, many fear that science is making this traditional Western-Civilization-View of man obsolete. Many across society believe that science "has set its own conditions and imposes its own values" on society, others warn that "society (is) dehumanizing rapidly," while still others claim that we are heading for "scientific control of society" (see endnotes 42, 43).[49,50,51,52] Independently of the validity of such claims, *the view from science, as presented by most scientists, clashes with the view upheld by a large fraction of society that man is the supreme value par excellence.*

And one can go even further, the fear of many in society of the possible effects on man of the recent scientific developments in biomedical sciences. They point to the ethics of human genetic engineering and to the possibility of inheritable genetic modifications in humans and thus they fear that humans are to be turned into and be bred like animals, hence signalling the end of man (see endnote 52). Independently of the validity and the extent of those fears, it is evident that t*he downfall of man unavoidably means the downfall of science.*

Fourth limit: Issues beyond the province of science.

Science is not the only way to the truth, and to claim otherwise, as many scientists do, is a distortion of science leading to conflict. The world is a hybrid of many things, and there are other, complementary, ways to the truth besides science such as those of art, philosophy and faith. Science deals with questions that can be defined scientifically, that can be studied scientifically, and that can have a chance to be answered scientifically; and scientists should demand respect by society of the proven scientific facts. However, science deals with neither ethical judgments, nor with the ultimate meaning of life. There are neither ontological experiments in science nor laws which describe our love and respect for each other. Those lay outside the province of science.

Beyond science, beyond the physical and the biological, beyond that which can be proved by the method of science and can be measured by the scientific instruments, lay the spiritual, the cultural and the intellectual traditions, the values of man, and the teleological concepts of philosophy and religion of which science does not speak. *To dismiss those "non-physical" aspects of human reality because they are not "proved" by the method of science, or to abandon science's metaphysical neutrality and transform science to a myth, limits the universal acceptance of science by society and presents the scientist antagonistic to traditional world-views.*

The general acceptability of science is largely because science makes no metaphysical claims. When that premise is abandoned, science will be judged differently and science will face a more confrontational society. *This will limit science's universality.*

Fifth limit: Perception of science as power to suppress and to destruct, and perception of scientists as instruments for negative use of scientific knowledge.

It is not possible to separate science from the consequences of the negative impact of the application of scientific knowledge. Increasingly, more people in society point to the dark side of the applications of science and picture science as a source of dangerous knowledge, which is used for destruction and limitation of man's freedom, safety and privacy. "The frightening thing which we did learn during the course of the war (WWII)", said I. Rabi "was how easy it is to kill people when you turn your mind to it. When you turn the resources of modern science to the problem of killing people, you realize how vulnerable they really are".[53,54] Science is thus looked at by a fraction of society as having set loose against society unimaginable forces capable of causing widespread destruction and suffering, be it through nuclear weapons, chemical agents, or biomaterials. Look, they say, at the dimensions of the nuclear arsenals (Table 3.1).[55] Today, the nine known nuclear powers have collectively over 20,000 nuclear warheads ready for immediate deployment; each of the thermonuclear bombs in these arsenals has a typical explosive power of several megatons. And they go further, they point out that the cataclysmic consequences of these weapons are with us because science and the scientists made it so. Because, since WWII, the frontiers of science and technology have become the frontiers of weaponry. Many of these weapons, including nuclear weapons, have been recommended, invented, developed and perfected by scientists and engineers, in the beginning and since. *The fear that science is increasingly becoming a power*

Table 3.1 The nuclear arsenal of the nine (9) countries known to have nuclear weapons[a]

Country	Date of first explosion	*Estimated* number of warheads
USA	1945	5400
Russia	1949 (USSR)	14,000
UK	1952	185
France	1960	<350
PRC	1964	<160
India	1974	100–140
Israel	1979?	100–200
Pakistan	1998	60
North Korea	2006	0–10

[a]Based on figures given by R. S. Norris and colleagues in several articles in the Bulletin of Atomic Scientists (see endnote 55); Also, see Reference 1 and S. Fetter et al., Physics Today, April 2018, pp. 33–39.

for suppression and destruction, and the perception of scientists as instruments of war diminishes their positive image and clouds their benevolent arguments.

Scientists have always had some part in military engineering (see endnote 1).[56] However, never before has so large a part of science been employed in this way than recently. It is estimated[57] that about half of the scientists and engineers in the USA and the world today are employed in the military field, many devising bigger and better weapons, weapons delivery systems, laser-guided bombs, military communications, as well as new nerve gases, germ warfare, so on. Science and scientists are unquestionably responsible for the dangerous nature of modern weapons – without modern science such weapons would not have been possible. The most horrifying possibility of modern war is of course nuclear war. Many scholars agree that the two superpowers have stockpiled enough nuclear weapons to destroy human life, and much of the rest as well, *many times over*. And alas! nuclear weapons technology is proliferating; new countries are engaged in efforts to acquire nuclear weapons[58] as we face another existential challenge, this time from bioscience.[59] What a distortion of science!

Interestingly, a 2011 editorial in *Nature* magazine[60] writes: "Twenty years after the end of the cold war, scientists and the military still need each other…. The US defense complex is the world's largest investor in military research. Much of the money has gone into developing weapons of unprecedented lethality, but large fraction supports 'dual-use' research, whose products – from the Internet to the Global Positioning System – have enriched society as a whole."

The atom bomb was the invention of scientists; they began the work on the atom bomb at their own volition. It is not usually the job of university professors to work on weapons of mass destruction, many have noted. Why then did these scientists initiate such work? Fear, they say, that the scientists of the other side could develop the bomb first and use it. And yet the nuclear bomb led to the nuclear power reactor, exemplifying the dangerous connection, **atoms for peace via atoms for war!** *There is thus a pressing need for radical scientific change, a need for a paradigm shift in the functions of modern scientists. Science will be severely limited, unless its power to suppress and to destruct, and to use scientists in this process, is curbed.*

Sixth limit: The careless scientist; scientists beyond the borders of science.

Scientists often step over the scientific norms, step over the borders of science, and become antagonistic in matters not scientific.[61] They speak on behalf of science on non-scientific matters or even on trans-scientific questions, spreading criticism, for instance, to the realm of religious belief, based not on what science says but rather on personal philosophy and personal world view. Geneticists tell us that they have discovered "The Language of God," theoretical physicists that they have discovered "The God Particle," astrophysicists that they have discovered "The Mind of God" and that their theory now says God does not exist ("God is unnecessary"), evolutionary biologists that they have discovered "The Selfish Gene" and the "God Gene" and still others that they have discovered the "Theory of Everything". Such expressions, even if they are not taken seriously, give the erroneous impression that science is omnipotent.

It is exceedingly disturbing to see scientists deal with God *scientifically!* To deify scientific theories is to turn science into a myth and expose science to undue criticism. Many in society are genuinely concerned that science is being turned into scientism and scientists into the High Priests of a new world view aiming to replace traditional world-views.

We need to adhere to the scientific tradition and to confine ourselves to questions which science can answer. We need to observe the proper boundaries of science. And we need to distinguish when we are doing science and when we are extrapolating from it, *especially when we are teaching students.*

Equally discomforting is the fact that scientists frequently abdicate their responsibility to the norms of science in favour of national and commercial interests. As the percentage of scientists who work for governments and industries increases, problems of freedom of inquiry and communication increase.

Even more troubling are the reported increased incidents of fraud in science,[62,63] suggesting that serious breaches of ethical behavior in scientific research are on the rise.[64] Typical serious breaches of ethical behavior that have appeared recently include those in South Korea[65] and those at premier USA Universities and Research Centers (Berkeley, Columbia, Harvard, Yale, Bell Labs, MIT).[66,67,68,69] Such behavior clearly weakens the bonds between science and society and lowers the trust of the scientist by society. Establishing the validity of each new scientific result is essential, for there is no scientist who did not err and because there have been examples of scientists who believed their results were valid but they were not (e.g., cold fusion[70,71]). There are also honest mistakes which are subsequently corrected and the respective original publication retracted. Such is the recent case involving a group of scientists at CERN in Italy, who in 2011 reported an experimental finding that neutrinos had traveled faster than the speed of light.[72,73] If confirmed, this result would have disproved Einstein's 1905 Special Theory of Relativity and would have contradicted more than a century of physics research based on the assumption that nothing exceeds the speed of light in vacuum. In making the announcement the leader of the research group urged caution, stating that the group had tried and failed to find a mistake in the research and that it was up to the scientific community to examine and replicate the work. The announcement was widely publicized. Subsequently, in 2012, experiments performed by a different group at the same laboratory found that neutrinos travel at the same speed as light. This is a sad story of honest research, which has been corrected through subsequent work. The story, however, raises questions as to when and how research groups and institutions should announce or publicize results that would be considered revolutionary or anomalous. Situations such as these differ from those where scientists were wrong and it has become clear that the scientists involved knew their results were fraudulent.

There is still much to be desired as well, in answering the cynical criticism levelled against us, namely, that "in the end, most scientists will do whatever there is money for doing." Regrettably, this has been shown to be the case time and again since Daedalus in ancient Greece.

There is, thus, a need to tame the arrogant and irresponsible scientist and to uphold the scientific values and tradition of an open-mind, modesty, honesty, and tolerance, and to improve our image. *The diminution of respect for and trust of the scientist by society limits the acceptance of science by society.*

What, then, can history tell us about the future of the "scientific" civilization? Clearly, any civilization's survival depends on its ability to adjust to change, to adapt. Science has no problem with that. Throughout history, civilizations have risen, reigned and fallen. For how long then might the scientific civilization continue to rise and reign, and when might it be expected to decline and ultimately collapse? The rise and reign of a civilization, it has been argued,[74] depend on its effective transmission to future generations, and one of the obstacles to this transmission is the subjective nature of its criteria. Science's "objectivity" must then be closely guarded if the scientific culture is to be effectively transmitted to future generations.

The universality of science requires coherent integration of science and its values into the world culture and this cannot be done through fear or scorn, or material promise, or through biological modification and manipulation of *Homo sapiens.*

3.6.2 Needs of the Universality of Science

For science to become truly universal:

- The trust of society in science and the scientist must be safeguarded and indeed it must be enhanced; the most significant factor in effecting this goal is the responsibility of the scientist.
- The scientific culture and the humanism inherent in science need to be more effectively communicated to society.
- The scientists should address the fears and concerns of society with modesty and must respect the dignity of man.
- The values of science and the traditional values of society need to achieve mutual accommodation.
- Science needs to work with society to address the ferocious problems facing humanity today – such as those of war and peace, deterioration of the environment, climatic change, and world poverty – for which a strong science is necessary but is not sufficient.
- Science must reassess its deep involvement with the machinery of war.

3.7 Science and Society

3.7.1 The Scientist

The crucial element of science has been and still is the scientist. Who is he? I attempted to answer this question in a book I published in 2001 and noted then, that today neither science is as distinct a term as it once was, nor is there distinctiveness in the term scientist. The proliferation in the numbers of those working in science today, the expansion of the scientific endeavor, and the vast uses and applications of scientific knowledge has resulted in more than one scientific identity. Today, the term scientist embraces many and diverse people with broad and heterogeneous areas of expertise, interests, attitudes, and values. In science, today, there are many science workers and relatively fewer scientists, and while both are indispensable elements of science, the heart of science remains the scientist.

Historically, scientists formed a community with no boundaries, where, in principle, everyone is free to enter, to work, to express his or her views, to be heard and to be contradicted. They adhered to the values inherent in the practice of science. Their community, though highly competitive, has traditionally been rather stable and largely incorruptible, sustained by a sense of dignity for its members. The scientific community itself has been largely shielded from social and metaphysical controversy by its limited impact on society, especially on man, and by metaphysical neutrality. Today, this is so no more. Hence the question: will the scientific community continue its stability and adherence to its tradition and norms so necessary for the benefit of both science and society?

Today, many young people seem not to be attracted by the challenges of a scientific career; many drift to easily-get-rich jobs rather than take the difficult road of becoming a scientist. The distinct characteristics and principles of science which are embedded in its tradition and method of inquiry need to be recognized and adhered to by the young scientists; they are normally learned tacitly in the execution of research and the proper guidance of the mentor-professor. There seems to be a need to broaden a scientist's education and societal perspective for there are difficult questions which are more frequently being asked today than in years past. Questions such as, could there be scientific knowledge the possession of which is harmful to society and thus further accumulation of such knowledge inappropriate? Is the primacy of unhindered right to new knowledge absolute, or should it be moderated by the legitimate concerns of society? *Whatever the answers to these and to other related questions might be, it is certain that man, as scientist and as citizen, will be increasingly limited by the burden of responsibility the power of scientific knowledge imposes on him.*[75]

3.7.2 Scientist and Society

3.7.2.1 Mutual Responsibility

Citizens of many countries around the world have different attitudes toward science and the scientists. Many citizens continue to hold scientists in high regard, while others mistrust them and at the same time expect miracles from them. It is the mutual responsibility of scientists and society to achieve a better understanding.

While today more countries invest in science and science education than previously, there is still a need to increase the science literacy of the citizen.[76] It is essential for society to recognize that virtually every major issue confronting society has a science and technology component requiring public understanding. It is also of utmost importance for society to appreciate the value of freedom in the execution of scientific research and the necessity to secure conditions conducive to scientific freedom. Only in a completely open society can the integrity of science be maintained and the dark side of science diminished. It is thus,

- The mutual responsibility of scientists and society to curb the power of science to suppress and destruct and to deploy scientists in this process. There is a need for radical change in this regard, a need for a paradigm shift in the functions of modern scientists. Science needs to reassess its deep involvement with the machinery of war.
- The mutual responsibility of scientists and society to predict, prevent and manage the risk against the idea of man associated with the progress of science. There will be immense future challenges to science and to human values arising from the impact of science and scientific technology on man.
- The mutual responsibility of scientists and society to require that the application of scientific knowledge is compatible with the values of society. For this, scientists and society must achieve accommodation between their mutual value systems and enhance their mutual trust. Obviously, the ethics of modern man cannot be based on science, but neither can it be separated from it, nor can science claim to be amoral.

3.7.2.2 Needs of Scientists and Society

Society needs to be more open and willing to embrace the acceptance of science. Society still fails to fully appreciate what science is providing for it despite the benefits it derives from science and although modern society will cease functioning without science and science-based technology. All too often society takes the benefits of scientific discovery for granted and all too often society exaggerates out of fear or ignorance potential negative impact and risk of new scientific knowledge. *Society, therefore, needs to accommodate science's unique ways of functioning and adjust to the facts of new scientific discovery, fully cognisant that scientific knowledge comes with benefits, but also with "peril and pain".*

Repeatedly in the recent past the relationship between the scientists and society has been strained by several key issues – embryonic stem cell research and climate change to name just two. In such instances, society is not just sensitive to what science does or does not do, but oftentimes society overreacts casting aside long-range benefits. The recent reaction regarding scientific data used to evaluate climate change[77] makes the point and may suggest that the trust between science and society is rather fragile. *An important aspect of the scientific literacy of society then should be enhancement in society's ability to recognize that while even in science errors are made, inherent in the method of science is the capability to discover and to correct such errors.*

Only in a free and open society can the integrity of science be maintained, and a free and open society is foremost society's responsibility.

3.7.3 The Scientist as Policy Advisor and as Advocate

Science advisors have been around for a very long time. One might in fact argue that the first scientific advisor in human history was Aristotle: he was the advisor of Alexander the Great. However, the emergence of scientists as political advisors, peacemakers and diplomats is a recent phenomenon, largely a product of the role played by scientists in WWII activities.

Today, enormous new scientific knowledge is generated across all fields of science, which is significant for human wellbeing; this powerful scientific knowledge is easily accessible and can quickly be put into practical use. Thus, the view is prevalent that scientists have a responsibility to advise governments, decision makers, and the public, of the possible benefits and risks of new scientific knowledge and science-based technology and to help them choose wisely between available options. There is a need to develop ways for "Science for Policy" activities, which will make possible the input of scientific evidence into the decision-making process and aid the resolution of social issues, claims and conflicts. Such an engagement of the scientist requires deep knowledge of the specific scientific issue and a holistic rather than a reductionist approach in translating scientific evidence into public policy; accountability in public policy, many have said, requires scientific evidence to be correctly embedded into the democratic process. The most important criteria for a scientist's contribution in this capacity are scientific competence, integrity, independence and transparency. These are basic prerequisites for an impartial assessment of the facts of science pertaining to the issue at hand worthy of the trust of society. To examine the impacts of policy decisions and to help mitigate their risks it is necessary to base such decisions both on science and on the values of society, to situate scientific evidence in the context of society's value system, and political judgments to be evidence-based. The subject has been discussed extensively for some time and a recent account can be found in Reference[78].

"Science for Policy" is needed to:

- *Aid society and decision makers in crises with scientific dimensions.* Examples of such crises are the Severe Acute Respiratory Syndrome (SARS) and other epidemics, the Fukushima nuclear accident, and earthquakes, tsunamis, hurricanes, floods, volcanic ash clouds, terrorism, and so on.
- *Clarify scientific claims on important controversial scientific-technological issues where answers are still not clear and claims are still not fully trusted.* Using the EU as an example, one can cite several science policy controversies of this kind: climate change, GM crops, fracking, food safety and security, water security, and a series of environmental and energy issues. And there are still science and technology relevant issues which may transcend the technical knowledge needed to provide a complete resolution of the issue, but scientific advice is

nonetheless sought to help the decision-making process even before a complete scientific understanding is reached. For instance, the possible effects of electromagnetic radiation on human health; although several major studies[79,80] have found that the use of mobile phones and exposure to radiation from high-voltage transmission lines does not pose any health risks, public concern about radiation from mobile phones and high-voltage transmission lines continues.

- *Delineate proposed claims for or against a given issue*. Invariably, risks arise when businesses or interest groups interpret scientific facts beyond the truth they contain. Often even honest science is mistrusted because of where it gets its funding. It is argued as well that even in countries where free speech is the norm, institutions are not always as open as necessary to dissenting voices. Impact assessments prepared to accompany policy proposals may, for example, be deliberately limited in the options they considered and in the sources of data they used, with the aim of achieving a desired outcome. Lack of openness in evidence gathering oftentimes has resulted in impact assessments being focused more on risks than on opportunities, in being more cautionary.
- *Choose wisely the mechanisms from which advice is gotten*. Today, it seems everyone wants to have scientific advice (especially the government) and everyone wants to give scientific advice, foremost to the government! Thus, debates over structures and procedures necessary for sound scientific advice abound. Unquestionably, society needs broad-based, open, evidence-gathering mechanisms to act. Four structures commonly used are: *individual scientists, chief scientific advisors, advisory councils/advisory committees, National Academies and Academy Organizations*. There has actually been a proliferation of Academy Organizations [International Council for Science (ICSU), InterAcademy Partnership (IAP), InterAcademy Medical Panel (IAMP), Federation of European Academies of Medicine (FEAM), European Academies Science Advisory Council (EASAC), All European Academies (ALLEA), European Council of Applied Sciences, Technology and Engineering (Euro-CASE), Academia Europaea (AE), and others][81] offering "independent" and "competent" scientific advice to governments and national and international organizations, "which often moderates extreme views on key issues and balances advocacy". National, regional, and global Science Advice Mechanisms are necessary for expert, competent and conditionally- and-contextually-independent advice.
- *Delineate the role of the scientist as a policy advisor and as an advocate*. The views of scientists (whether acting alone or as members of academies/organizations/committees) are respected because they are supposed to be objective and independent experts in the field advice is sought, but when they act as advocates they are likely to be in conflict with the professional norms of science. Advocacy by the scientists themselves on behalf of any issue be it the environment, global warming, shale gas extraction, GMFs, stem cells, or synthetic biology, may be a real or perceived attempt to affect the opinions of the general public or certain groups of population, or the decision making of politicians, legislators and governments. It is obviously not just risky but unfair to assume that when scientists become advocates they become partisans and are no longer "neutral conveyors of scientific information". Yet, scientific advice almost always contains shades of

opinion not entirely scientific, and the goals and methods of advocacy can be in conflict with the goals and methods of science. It is not uncommon to find the experts whose advice is sought to be the ones with vested interests in the subject they are called upon to give advice. Thus, the advice should be sought more broadly and should include competent people from outside the subject matter concerned. Does everything that scientists say or advocate have the backing of science and is self-regulation of scientists adequate to handle the pressures placed upon their scientific integrity? While potential sources of conflict will always accompany science advising, one aspect is clear: *support good, evidence-based policy, and clarify the boundaries and validity of the information provided.*

Undoubtedly, we are witnessing new paradigm shifts as to the role of scientists and their scientific societies.

References and Notes

1. Loucas G. Christophorou, *Place of Science in a World of Values and Facts*, Kluwer Academic/Plenum Publishers, New York 2001.
2. L. G. Christophorou, *Atomic and Molecular Radiation Physics*, Wiley-Interscience, New York 1971.
3. L. G. Christophorou (Ed.), *Electron-Molecule Interactions and Their Applications*, Vols. 1 and 2, Academic Press, Inc., New York 1984.
4. Loucas G. Christophorou and James K. Olthoff, *Electron Interactions with Excited Atoms and Molecules*, Advances in Atomic, Molecular and Optical Physics, Vol. **44**, 2000, pp. 155–293.
5. Loucas G. Christophorou and James K. Olthoff, *Fundamental Electron Interactions with Plasma Processing Gases*, Kluwer Academic/Plenum Publishers, New York 2004.
6. Loucas G. Christophorou, *Energy and Civilization*, Academy of Athens, Athens 2011.
7. Λουκάς Γ. Χριστοφόρου, *Ενέργεια: Επιστημονική, Φιλοσοφική και Θεολογική Διάσταση*, Πρακτικά της Ακαδημίας Αθηνών, τ. **89** Α′, σελ. 27–48, Αθήνα 2014. Loucas G. Christophorou, *Energy: Scientific, Philosophical and Theological Dimension*, Proceedings of the Academy of Athens, Vol. **89** A′, pp. 27–48.
8. All dates quoted are those given in Wikipedia.
9. Λουκάς Γ. Χριστοφόρου, *Αέναη και Κρίσιμη Αλλαγή*, Πρακτικά Ακαδημίας Αθηνών, τ. **84** Α′, σελ. 109–132, Αθήνα 2009. Loucas G. Christophorou, *Perennial and Critical Change*, Proceedings of the Academy of Athens, Vol. **84** A′, pp. 109–132.
10. Kurt Gödel cautioned (see endnote 11) that "No finite set of axioms and rules of inference can ever encompass the whole of mathematics. Given any finite set

of axioms, meaningful mathematical questions can be found which the axioms leave unanswered." See also Keith Devlin, Science **298**, 6 December 2002, pp. 1899–1900.
11. http://plato.stanford.edu/entries/goedel/.
12. See, for instance, Phillip Anderson, Science **177**, 4 August 1972, pp. 393–396; Paul Davies, *The Cosmic Blueprint*, Templeton Foundation Press, Pennsylvania 1988; E. O. Wilson, *Consilience—The Unity of Knowledge*, Vintage Books, New York 1999.
13. Loucas G. Christophorou, *The Universality of Science: Limits and Needs*, invited lecture at the annual ICSU European Members Meeting, 29–30 September 2009, Podgorica, Montenegro; Loucas G. Christophorou, invited lecture entitled *The Universality of Science: Limits and Needs*, at a special meeting on *Science Literacy and the Responsibility of Scientists* organized by the EU members of the International Council for Science (ICSU) in connection with the ICSU General Assembly meeting in Rome (September 2011); Rendiconti Lincei, Vol. **23**, Supplement 1, September 2012, pp. S23–S27.
14. Steven Weinberg, *Dreams of a Final Theory*, Vintage Books, New York 1994, Chapters 1 and 2.
15. Λουκάς Γ. Χριστοφόρου, *Η Επαγωγική Μέθοδος της Φυσικής Επιστήμης (Από τα Μόρια στον Άνθρωπο;)*, Πρακτικά της Ακαδημίας Αθηνών, τ. **82** Α′, σελ.1–30, Αθήνα 2007. Loucas G. Christophorou, *The Inductive Method of Physical Science (From Molecules to Man?)*, Proceedings of the Academy of Athens, Vol. **82** A′, pp. 1–30 Athens, 2007.
16. Robert John Russell, in *Physics, Philosophy and Theology*, Robert John Russell, William R. Stoeger, J. J., and George V. Coyne, S. J. (Eds.) (Third edition), Vatican Observatory-Vatican City State 1997, pp. 343–374.
17. See Reference 15 and sources cited therein.
18. Phillip Anderson, *More is Different*, Science **177**, 4 August 1972, pp. 393–396.
19. Paul Davies, *The Cosmic Blueprint*, Templeton Foundation Press, Pennsylvania 1988.
20. http://www.chem1.com/acad/sci/aboutwater.html.
21. Edward O. Wilson, *Consilience—The Unity of Knowledge*, Vintage Books, New York 1999.
22. P. Davies, *The Mind of God*, Simon and Schuster, New York 1993.
23. E. P. Wigner, *Physics and the Explanation of Life*, Foundations of Physics, Vol. 1, No.1, 1970, pp. 35–45.
24. Stuart A. Kauffman, *Reinventing the Sacred*, Basic Books, New York 2008.
25. Stuart A. Kauffman, in M. M. Waldrop (Ed.), *Complexity*, Simon and Schuster, New York 1992, p. 122.
26. Bas C. van Fraassen, in J. Hilgevoord (Ed.), *Physics and Our View of the World*, Cambridge University Press, Cambridge 1994, p. 114.
27. Bas C. Van Fraassen, *The Empirical Stance*, Yale University Press, New Haven 2002.

28. Michael Polanyi, *Personal Knowledge—Towards a Post-Critical Philosophy*, The University of Chicago Press, Chicago 1958; Michael Polanyi, *The Study of Man*, The University of Chicago Press, Chicago1959.
29. Thomas S. Kuhn, *The Structure of Scientific Revolutions*, (second edition), The University of Chicago Press, Chicago 1970.
30. Niels Bohr, *The Philosophical Writings of Niels Bohr—Essays 1933-1957 on Atomic Physics and Human Knowledge*, Vol. II, Ox Bow Press, Woodbridge, Connecticut 1987, pp. 32–66.
31. Martin Rees, *Before the Beginning*, Basic Books, New York 1998.
32. Nancy Cartwright, *The Dappled World—A Study of the Boundaries of Science*, Cambridge University Press, Cambridge 1999.
33. John D. Barrow, *The Constants of Nature*, Vintage Books, New York 2002; John D. Barrow and Joseph Silk, *The Left Hand of Creation—The Origin and Evolution of the Expanding Universe*, Oxford University Press, New York 1983.
34. For instance, the intensity of the Coulomb field varies as $1/r^2$, where r is the distance from the charged particle. Similarly, the intensity of the gravitational field varies as $1/r^2$, where r is the distance from the mass generating the field.
35. The gravitational and electrical potential energy differ enormously in their relative strengths. For any distance of separation r of two bodies, the gravitational potential energy is almost 10^{39} times smaller in magnitude than the electrical potential energy of two charged bodies at the same distance apart.
36. B. P. Abbott et al. Phys. Rev. Letts **116**, 061102-1-16, 12 February 2016; Sung Chang, Physics Today, 14 April 2016, pp. 14–16.
37. M. Coleman Miller, Nature **531**, 3 March 2016, pp. 40–42; Adrian Cho, Science **351**, 19 February 2016, pp. 796–797.
38. See, for instance, R. P. Crease and C. Westfall, Physics Today, May 2016, pp. 30–36.
39. Arnold S. Relman, "The New Medical-Industrial Complex", The New England Journal of Medicine, Vol. **303** (No. 17), 1980, pp. 963–970.
40. Attention is also being drawn to the megatrends (large economic units such as Google and Apple) seen in the Silicon Valley "that exercise a form of sovereignty of their own" (Sidney D. Drell & George P. Shultz (Eds.), *Andrei Sakharov—The Conscience of Humanity*, Hoover Institution Press, Stanford University, Stanford, California 2015, p. 135).
41. Another area of concern is the relationship between security and science and the difficulty in balancing scientific freedom and national security concerns. Kraemer and Gostin (John D. Kraemer and Lawrence O. Gostin, *The Limits of Government Regulation of Science*, Science **335**, 2 March 2012, pp. 1047–1049) focused on this issue referring to the case of genetically modified H5N1 avian influenza viruses. They write that the US National Science Advisory Board for Biosecurity, "recommended that two journals, Science and Nature, retract key information before publication", because of "concerns that published details about the paper's methodology and results could become a blueprint for bioterrorism". Researchers want no restriction and access, no

government intrusion.... "However, security advocates believe the greater risk is that mutated virus could escape or that knowledge about these mutations could get into the wrong hands. They suggest that research of this kind should not be funded or undertaken in the first place."

42. D. A. Schon, *Beyond the Stable State*, Random House, New York 1971, p. 22.
43. Havel, V., *The Need for Transcendence in the Post-modern World*, Speech in Independence Hall, Philadelphia, July 4, 1994; http://www.worldtrans.org/whole/havelspeech.html.
44. Jacques Maritain, quoted by Gerald Holton in Rosemary Chalk (Ed.), *Science, Technology, and Society—Emerging Relationships*, The American Association for the Advancement Science, Washington, DC 1988, p. 50.
45. InterAcademy Council (IAC), *Responsible Conduct in the Global Research Enterprise*, A Policy Report, September 2012 (ISBN: 978-90-6984-645-3).
46. Some countries despite the high quality of their scientists have very low percentage of their GDP allocated for support of search (Greece, for example, allocates < 0.6 % of the country's GDP).
47. National Science Board (NSB), *Science and Engineering Indicators*, 2012, Arlington, VA.
48. A. M. Weinberg, Minerva **10**, April 1972, p. 209; *Science* **177**, 27 July 1972, p. 211.
49. Rosemary Chalk (Ed.), *Science, Technology, and Society—Emerging Relationships*, The American Association for the Advancement of Science 1988, p. 48 and p. 53.
50. Loucas G. Christophorou and George Contopoulos (Eds.), *Universal Values*, Academy of Athens, Athens 2004 (ISBN: 960-404-061-8).
51. Jérôme Bindé (Ed.), *The Future of Values—21st-Century Talks*, UNESCO Publishing, Paris/Berghahn Books, New York 2004.
52. See, for instance, Leon R. Kass, *Life, Liberty and the Defence of Dignity—The Challenge for Bioethics*, Encounter Books, San Francisco 2002; Gregory Stock, *Redesigning Humans—Choosing our Genes, Changing our Future*, Houghton Mifflin Company, Boston 2003; C. S. Lewis, *The Abolition of Man*, Harper, San Francisco 2001.
53. I. I. Rabi, *Science, The Center of Culture, Perspectives in Humanism*, The World Publishing Co., New York 1970, p. 71.
54. Increasingly society considers that scientists have a moral responsibility to answer for what they do in their research. For instance, no experiments can be defended aiming to determine the most economical and best engineered way to carry out the mass destruction of people.
55. R. S. Norris and collaborators in the Bulletin of Atomic Scientists 2002, 2005–2008 (see also References 1 and 9).
56. Classic examples of military engineers are Archimedes, Leonardo da Vinci and Michelangelo. It should be noted that Abraham Lincoln established the US National Academy of Sciences during the Civil War in part to provide advice on military matters.

57. Jeff Schmidt, *Disciplined Minds*, Rowan and Littlefield Publishers, New York 2000.
58. Sidney D. Drell & George P. Shultz (Eds.), *Andrei Sakharov—The Conscience of Humanity*, Hoover Institution Press, Stanford University, Stanford, California 2015.
59. L. F. Cavalieri (The Bulletin of the Atomic Scientists, December 1982, p. 72) writes: "There is a striking similarity between nuclear science and genetic engineering. Both major accomplishments confer a power on humans for which they are psychologically and morally unprepared."
60. Editorials, Nature **477**, 22 September 2011, p. 369.
61. See, for instance, Jian Hilgevoord (Ed.), *Physics and our View of the World*, Cambridge University Press 1994; M. Ruse, *Is Evolution a Secular Religion?* Science **299**, 7 March 2003, p. 1523; Michael Ruse, *The Evolution-Creation Struggle*, Harvard University Press, Cambridge, Massachusetts 2005; Richard Dawkins, *The God Delusion*, Houghton Mifflin, Boston 2006; Elaine Howard Ecklund, *Science vs. Religion—What Scientists Really Think*, Oxford University Press, New York 2010.
62. Adil E. Shamoo and David B. Resnik, *Responsible Conduct of Research*, Oxford University Press, New York, 2003.
63. E. Garafoli, Rendiconti Lincei, September 2015, Vol. **26**, Issue 3, pp. 369–382.
64. While there may well be increased rates of plagiarism today, this may in part be due to the increased access to the Internet and the methods being applied to define and to detect plagiarism.
65. D. Normile, G. Vogel, and J. Couzin, Science **311**, 13 January 200, pp. 156–157; M. K. Cho, G. Magnus, Science **311**, 3 February 2006, pp. 614–165; D. Normile and G. Vogel, Science **310**, 16 December 2005, pp. 1748–1749.
66. New York Times, August 28, 2010 (Harvard researcher may have fabricated data in monkey study).
67. Science **298**, 4 October 2002, p. 30.
68. Robert F. Service, Science **296**, 24 May 2002, pp. 1376–1377.
69. Robert F. Service, Science **299**, 3 January 2003, p. 31.
70. Robert Pool, Science **243**, 31 March 1989, pp. 1661–1662; Science **244**, 7 April 1989, pp. 27–29.
71. Frank Close, *Too Hot to Handle— the Race for Cold Fusion*, Princeton University Press, Princeton, New Jersey 1991.
72. https://en.wikipedia.org/wiki/Faster-than-light_neutrino_anomaly.
73. A. Cho, Science **333**, 30 September 2011, p. 1809; A. Cho, **334**, 2 December 2011, pp. 1200–1201.
74. M. J. Moravcsik, Bulletin of Atomic Scientists, Vol. **29**, March 1973, pp. 25–28.
75. A recent Interacademy partnership report (Interacademy partnership, *Doing Global Science*, Princeton University Press, Princeton 2016) expresses concerns about the increase incidents of misconduct in research; it states: "irresponsible behavior and poor practices pose threads to the global research enterprise, could impair its effective functioning, and could even damage the

broader credibility of science." A similar publication by ICSU in 2008 provided "guidance about the responsibilities and freedoms of researchers", and in 2009, NAS-NAE-IOM (The US National Academies of Sciences, Engineering and Medicine) issued an educational guide "*On Being a Scientist: A Guide to Responsible Contact of Research*". In 2012 a publication by the InterAcademy Council (IAC) and the IAP (the Global Network of Science Academies) entitled *Responsible Conduct in Global Research Enterprise: A Policy Report*, describes the values of research and how those values should guide the conduct of research. The report acknowledges that "while different disciplines and countries have varying research traditions and cultures, the fundamental values of research transcend disciplinary and national boundaries and form the basis for principles of conduct that govern all research". In March 2017, ALLEA issued a revised edition of its earlier document entitled "*The European Code of Conduct for Research Integrity*".

76. Rendiconti Lincei, Vol. **23**, Supplement 1, *Science Literacy*, September 2012.
77. See comments by Martin Rees and Jane Lubchenco in Science **327**, 26 March 2010, pp. 1591–1592. See, also, R. J. Cicerone, editorial, Science **327**, 5 February 2010, p. 624; Eli Kintisch, Science **327**, 26 February 2010, p. 1070.
78. James Wilsdon and Robert Doubleday (Eds.), *Need Foresight for Scientific Advice in Europe*, Centre for Science and Policy, April 2015, Cambridge (ISBN: 978-0-9932818-0-8).
79. NIEHS Working Group Report, *Assessment of Health Effects from the Exposure to Power-Line Frequency Electric and Magnetic Fields*, National Institute of Environmental Health Sciences of the National Institutes of Health, US NIH Publication No. 98-3981, August 1998; for subsequent articles and reports on this subject see British Journal of Cancer, American Journal of Epidemiology, and Epidemiology.
80. On the question of mobile phone radiation and health, articles can be found in the journal of Health Physics, Journal of Exposure Analysis and Environmental Epidemiology, Occupational and Environmental Medicine, American Journal of Epidemiology and the British Medical Journal. See also http://en.wikipedia.org/wiki/Mobile_phone_radiation_and_health.
81. Actually, since 2016, five European Academy Organizations—Academia Europaea, ALLEA, EASAC, FEAM and Euro-CASE—are working together under the name SAPEA (Science Advice for Policy by the European Academies) to provide as a group "interdisciplinary evidence-based advice, reviews and other scientific input to the Science Advice Mechanism (SAM) of the European Commission".

Chapter 4
Scientific and Technological Frontiers

4.1 Introduction

No one really knows what science is doing today. What we do know is that increasingly scientific questions are becoming more difficult, in need of new scientific instruments, methods and facilities; better information technology and advanced computation[1]; new concepts and constructs and new mathematics to enable better understanding of higher levels of abstraction in basic science, and of complex systems, foremost biological; new initiatives to expand the scientific frontier further into space and time at both ends of the time and space scales. How then can anyone predict what would be frontier science and scientific technology in the future, say, by the end of this century and even beyond? Despite the difficulty, attempts have been made.[2,3,4,5,6,7,8]

Science will continue the exploration of the very small entities and the very big parts of the cosmos. It will penetrate deeper into the mysteries of elementary particles, "exoparticles", gravitational-wave astronomy,[9] exoplanetary astronomy, dark matter, and the beginning of the universe. Science will also face the challenge of complexity and emergence seeking new laws for understanding the behavior and the properties of complex systems, animate and inanimate; biomedical sciences will accelerate their move toward the atomic, the molecular, and the nanoparticle level of matter, and the understanding at this level will truly become a new frontier of excitement and of many science-based technologies.

Science-based technologies will continue to converge on three broad fronts: the manipulation of atoms (nanotechnology), the manipulation of genes (biotechnology/genome technology), and the manipulation of "bits" (information and communication technologies). Technologies in these areas will move downward to the manipulation of the very small and upward toward the emerging properties of the big and complex structures of animate and inanimate matter. The scientific and technological push may face fundamental limits; for instance, in the speed with which information can be transmitted from one point to another and in the

L. G. Christophorou, *Emerging Dynamics: Science, Energy, Society and Values*,
https://doi.org/10.1007/978-3-319-90713-0_4

amount of information which can be stored or processed. These and other frontier science-based technologies will continue to depend on the availability of energy and its appropriately-conditioned forms. Likely, they will all have *dual – positive and negative – aspects in their impact on society and on man himself.*

In this Chapter, we focus on the following four areas: (i) Complexity, (ii) Atomic and molecular biosciences (foremost molecular genetics and molecular medicine), (iii) Energy (possible new energy sources, energy carriers and useful transformations of energy), and (iv) New materials (nano-, bio- and info-materials and superconductors).

4.2 Scientific and Technological Frontiers

4.2.1 Complexity

4.2.1.1 Complexity in Nature

Complexity is an intrinsic property of matter; it is a consequence of its innate tendency to self-organize via the incredibly large possibilities of combinational interactions between its parts and their interactions with the environment. As the organization of matter grows more complex, new collective properties emerge, characteristic of their level of complexity. The trend toward increased complexity leads to increased diversity and the emergence of new combinational structures with new emergent properties (see Chap. 3).

Let us look at just two examples: one "simple" and the other "very complex". The first example is the water molecule (H_2O), to which we referred to earlier in Chap. 3. The water molecule is made up of three atoms, two hydrogen atoms (H) and one oxygen atom (O) (Fig. 4.1).[10] The properties of the water molecule depend on its quantum mechanical structure determined by the atoms H and O, especially the orbitals occupied by their outermost electrons. The water molecule however, has its own characteristic physical and chemical properties, for instance polarity; the polarity of the water molecule is due to its constituent atoms and their stereochemical arrangement, but it is not a property of the atoms themselves.

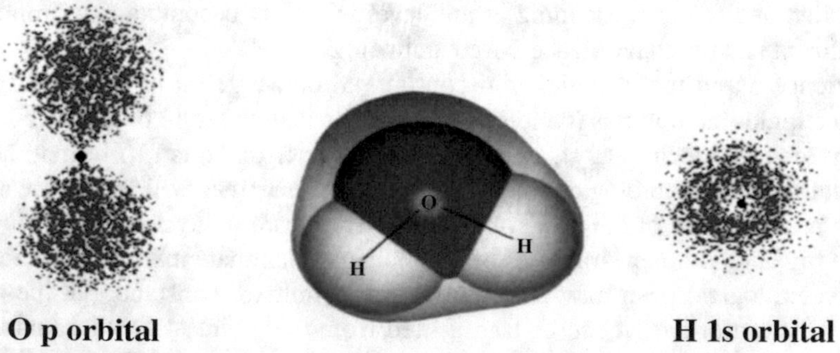

Fig. 4.1 The water molecule (see endnote 10)

As discussed in Sect. 3.2, the geometrical and electronic structure of the water molecule determines the way the water molecule interacts with other molecules. Water molecules form hydrogen bonds with other water molecules; water molecules form metastable clusters consisting of specific numbers of water molecules which are continually formed, broken up and reformed again; water molecules form complexes around positive and negative ions or electrons. Liquid water acquires a new property – *fluidity* – that no isolated water molecules possess. If, furthermore, liquid water is heated up to 100 °C, the very same water molecules "vaporize", the system undergoes a phase transition and becomes vapor; if the vapor is cooled down to 0 °C, the system undergoes another phase transition, the water molecules stop moving chaotically, take regular positions and arrange themselves in a hexagonal crystal structure known as ice. These changes, the new states of matter to which they lead and their emergent properties, have no meaning for the isolated water molecules; they are characteristic of the system as a whole.

The second example is the human brain, a very complex and a very complicated system indeed (Fig. 4.2).[11] Many scientists have drawn attention to the complexity of the human brain. As it is presently understood, the human brain contains ten billion neurons, each of which sends feelers or axons to link to about one thousand others. These connections play a role in creating our thoughts and memories. How this is done is still not known. What is known is that the number of possible thoughts or ideas the human brain can conceive is incredibly large (see endnote 11)[12]; according to some estimates[13]: $10^{70,000,000,000,000}$. This is an estimate of the number of different electrical patterns the human brain can hold. This number is truly horrendous compared, say, to the number of atoms in the observable universe (a mere 10^{80}) or the atoms the human brain itself consists of (merely 10^{27} atoms). The enormous number of the different electrical patterns the human brain can hold results from the complexity of the connections between its component parts. Complexity arises from the number of diverse ways in which the component parts can be connected, rather than out of the number of those component parts.

Unquestionably, the human brain is one of the most challenging present and future scientific frontiers. At the 2016 March meeting of the American Physical Society (APS), a call was made[14] for physics expertise to enter the field of brain research and neuroscience. The challenges pointed out at this meeting stretch from "modeling the biomechanics of the brain development, improving neuroimaging techniques, processing and analyzing the data from studies using these techniques", to quantifying the brain's material properties which "give the part of the brain known as the cortex its complex folded shape" that might allow an understanding of cortical folding. According to geneticist Miyoung Chun of the Kalvi Foundation, Popkin writes (see endnote 14), "the action today involves technologies that record the activity of single neurons, potentially allowing researchers to map out the entire brain circuits and explore the brain's computational code".

The workings of the brain, the "brain-mind problem", and the scientific interventions aiming in understanding and improving the brain's functions are but a few of the issues related to brain research and the associated ethics. If the structure

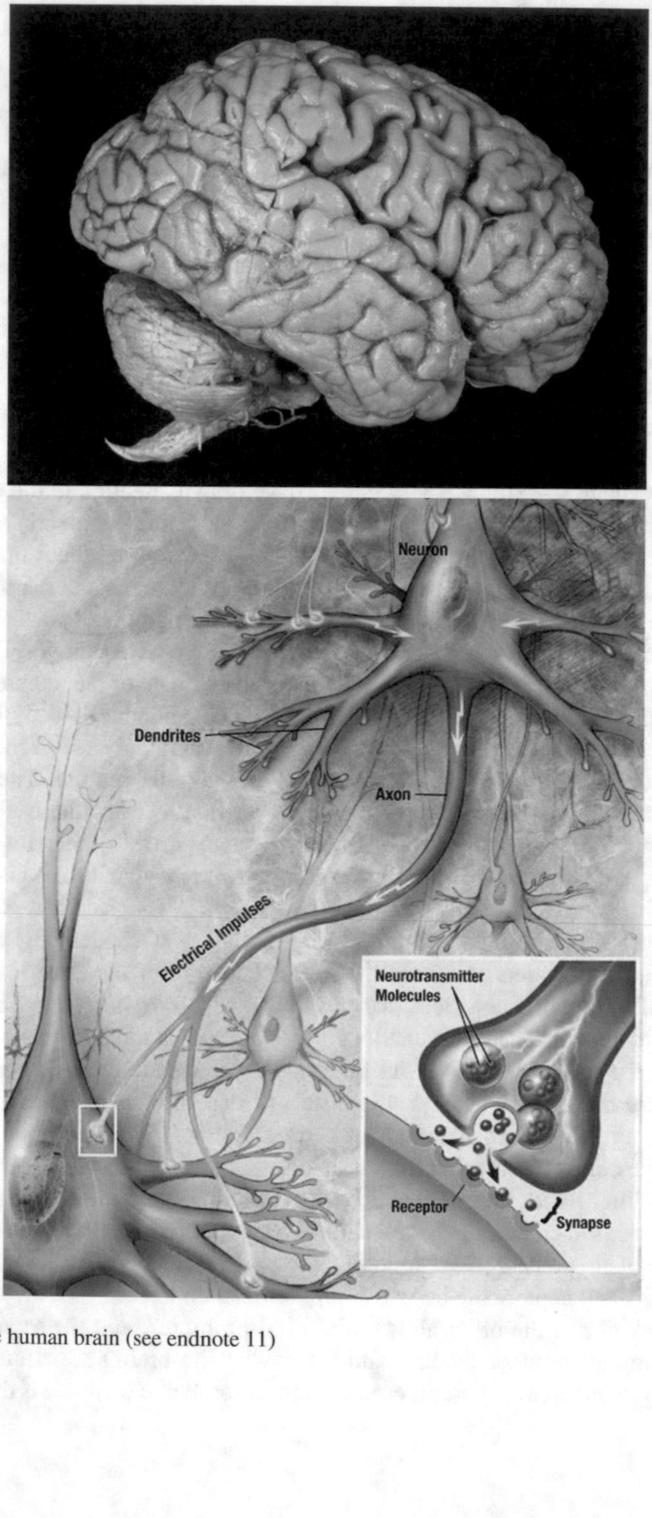

Fig. 4.2 The human brain (see endnote 11)

and function of the human brain is what determines who we are, as many assert, should we improve or manipulate the brain to our benefit? Clearly, we are entering neuroethics.

Complex systems may become unstable beyond a given level of size and complexity. Even at the microscopic scale this is so. Heavy nuclei, for instance, lose their stability and fission on their own when they become too large; complex molecules, as a rule, break up more easily than simple ones. Is there, then, an inherent limit to the complexity of every system, the human brain included? We know that as systems become more complex they require more energy to function. Do, then, systems break up because their complexity demands copious amounts of energy, which are themselves inefficiently used? Can ways be found to reduce a system's complexity before it reaches criticality, for instance, by enabling the system to function with smaller amounts of energy which are more efficiently utilized? Such questions may become easier to answer if complexity is governed by laws that can be discovered and can form the basis of a fundamental understanding of complexity itself and its internal workings, and if prediction can be made of the behavior of complex systems based on such laws[15] (see Sect. 3.2). What are then the new concepts and constructs that are necessary for the formulation of such laws? This remains a challenge for the future.

4.2.1.2 Complexity in Society and Values

Human society, history tells us, is increasingly moving toward higher levels of complexity: larger settlements supported by increasingly more complicated infrastructures; more institutions, more social needs and more specialization; larger information and communications loads and more societal interconnections through an elaborate web of systems and technologies. Society increasingly becomes more dependent on powerful technologies to support the services demanded by its population traditional needs and new habits such as the explosive growth in consumer, business, and government e-services. The cost of maintaining this societal complexity is increasingly becoming more difficult to afford principally because it requires: (i) processing enormous amounts of energy in an increasingly less efficient manner, and (ii) technological infrastructures which grow increasingly more complex and become increasingly more difficult to understand and to control.

Societal complexity and its maintenance, it is argued, destabilize society's institutions and diminish their adaptive capacity; they make society operationally vulnerable. In his book, entitled "The Collapse of Complex Societies", Tainter[16] argues that societies can reach and pass a point of diminishing marginal returns to investment in societal complexity. Once complex societies become unable to support their complexity, they crumble and unavoidably collapse. Yet, all indications are that present complex societies will get more complex in the future. They will thus require more efficient infrastructure, innovative technology, new information processing systems, and more energy.

The increase in societal complexity and the concomitant increases in human interactions and contacts – real and virtual – are accompanied by changes in human behavior. New types of human relations emerge, which bring along new challenges to traditional human values and ethics. For instance, human reciprocity weakens, and as it does, it weakens the effectiveness of the "the golden rule". Local and regional human problems and events become instantly "panhuman" stretching ethics in time and distance; ethics assumes new time- and space-characteristics, it becomes delocalized.[17] Will, then, the spectrum over which value judgment is expected to apply become too large for any value to be meaningfully effective? Is societal complexity a challenge to values?

In a similar fashion, the ethics of energy and the environment transcends locality demanding responsible global action stretched in space and time. Similarly, the new ethics of the perception of risk vs. benefit demands stretching in space and time. Nuclear waste is a case in point; there is a large divergence between the views of many scientists who view radioactive isotopes as dangerous but well understood and readily managed,[18] and members of the public who view nuclear waste as a long-lasting malevolent legacy from nuclear weapons and power reactors. The relevant ethical question is: should we be concerned about the far-out possibility that a nuclear-waste-disposal site may begin to leak, say, 100,000 years from now? May be, but certainly this is a new question to ask and it does not matter that such a question might be irrelevant to life today. It seems that citizens, especially in advanced societies, have become more risk averse, too familiar with and more willing to live with risk compared to previous generations.[19] A familiar other example is the risk of a nuclear war. While this danger hangs over present and future generations, in general, we do not talk much about it. We have familiarized ourselves with this high-level risk, and, in some way, we have adapted to it!

Adaptability, it has been said, is an asset for survival. Paradoxically though, the greatest threat to the quality of life and to human survival might be *Homo sapiens'* immense adaptability. How, one might ask, would we adapt to machines interacting with each other as algorithms, trading among themselves with little human involvement? The trend towards more interconnected and less comprehensible technological systems represents a gradual evolution of the technological support system of modern society towards levels of higher complexity, and thus toward higher levels of societal risk and instability. Science and technology by facilitating higher levels of societal complexity, challenge society.

4.2.2 Molecular Genetics and Molecular Medicine

We have discussed this subject in Chap. 2, and here we will recapitulate just a few aspects of that discussion for completeness.

In the previous century, we have witnessed the merger of chemistry with physics and gradually the merger of biology with both chemistry and physics. By the end of the twentieth century we have begun to see the gradual reduction of parts of medicine to atoms, molecules and genes, and the beginning of the remarkable explosion of molecular and genomic medicine, driven in part, by *bioinformatics*. Basic elements of these emerging technologies are the next generation of genome sequencing, genetic engineering, and big-data-driven medicine. In the manipulation of the very small lies new fundamental knowledge for understanding the behavior of the very big, which, undoubtedly, will lead to new technological frontiers in biology, medicine, bio- and nano- technology.

Examples of the new frontiers in these fields are the following:

- *Molecular and genetic roots of cancer (multigene damage),*
- *Stem cell technology,*
- *Designer genes,*
- *Germline gene modification,*
- *Synthetic biology,*
- *Synthetic viruses,*
- *Epigenetics,*
- *Human genetics,*
- *Prosthetics,*
- *Genetic modifications of plants and animals (Genetically Modified Organisms, GMOs).*

4.2.3 New Materials

Frontier science-based technologies will rely heavily on new materials. By way of example, let us look at just two categories of materials: *nanomaterials and superconductors*.

4.2.3.1 Nanomaterials

Nanomaterials are substances with dimensions less than ~100 nanometers (nm) (1 nm = 10^{-9} m). At these sizes, materials exhibit size-dependent properties. Nanomaterials are increasingly being used in bioscience, nanobiology, information science and technology, energy generation and storage, bio-physico-chemical processing and catalysis, medicine, and so on (see endnote 19). Nanomaterials research is rapidly expanding in the use of nanoparticles in medicine and cancer therapy. Also, nanomaterials and nanodevices are envisioned revolutionizing medicine whether through nanomachines or molecular robots.

Another important application of nanomaterials is in *nanophotonics,* the study of the interaction of light at the nanometer scale, which allows understanding of the flow of light at length scales far below the optical wavelengths. Many new developments have resulted, and further developments are expected to result, from increased control over the flow of light at length scales smaller than the photon wavelength in solid-state-lighting and in solar energy technologies. Generally, as photons are "shrunk" to nanoscale dimensions – ultimately approaching the scale of the wave functions of the bound electrons in atoms – new fundamental insights into the interaction of light with matter at sub-wavelength scales are expected, and light-based quantum technologies are envisioned driving forward the quantum communications information revolution.[20] Quantum states can share entanglement between several systems and these can encode information which is shared between these systems (see endnote 20)[21,22,23]; quantum light is an ideal medium for transmitting quantum information.

Nanomaterials will also play an important role in other new fields such as nanobiology and nanobiotechnology, which arise from the merging of biological science with nanoscience and nanotechnology.

Another related class of materials is *metamaterials and hybrid nanophotonics.* Metamaterials are artificial materials with an unusual optical response, formed by ordered or disordered collections of resonant nanoscale plasmonic scattering elements, and hybrid nanophotonics involve the simultaneous control of tightly confined light and phonons, electrons, spins and/or excitons interacting with light (see endnote 20).[24] Plasmonic nanostructures offer unprecedented level of light concentration and new perspectives to interface light and matter. In these and other advanced technologies, the crucial role of light and slow-electron pulses will continue to be at the basis of many future advances in both science and technology.

4.2.3.2 High-Temperature Superconductors

The electrical resistance of metallic conductors such as silver and copper decreases smoothly as the temperature (T) of the material is decreased toward absolute zero. There are, however, materials (metals, alloys, ceramics, organics), whose resistance abruptly goes to zero when the material is cooled below a certain (critical) low temperature, T_c; below that critical temperature T_c the material becomes a superconductor, able to maintain a current with no applied voltage.

A superconductor is said to be *low temperature* if its T_c is lower than that of liquid nitrogen (nitrogen boils at 77 K = −195.79 °C) and *high temperature* if its $T_c > 77$ K. In the former case, the superconducting state is reached by cooling the material using liquid helium (helium boils at ~ 4 K = −269 °C) and in the latter case using liquid nitrogen.[25]

Figure 4.3 shows the critical temperature T_c of various materials as a function of the year it was discovered and measured.[26,27] The highest temperature superconducting materials known today are the cuprates, which have demonstrated superconductivity at atmospheric pressure at temperatures as high as −135 °C (138 K) (Fig. 4.3). In 2015, hydrogen sulfide (H_2S) under extremely high pressure (~150 gigapascals, ~1.5 million atmospheres) was reported[28,29,30] to undergo superconducting transition at ~ 203 K (−70 °C). Subsequent studies (see Chang[31]) have shown that the superconductor is actually H_3S (and not H_2S) produced by pressure-induced dissociation of H_2S.

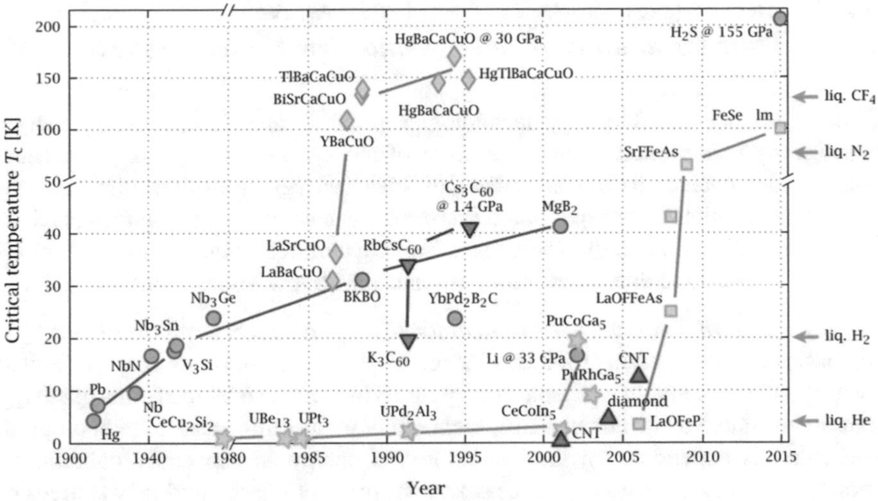

Fig. 4.3 Superconductivity from 1900 to 2015: Critical temperature T_c (K) vs. the year it was discovered and measured (see endnotes 26, 27)

A *room-temperature superconductor* is a material which would exhibit superconductivity at 0 °C (273.15 K). Although this is not strictly speaking room temperature (~ 20–25 °C), it is the temperature at which ice forms and it can easily be reached and maintained. Finding a room-temperature superconductor would allow creation of huge magnetic fields that require little power and would be of enormous multiple technological significance: for instance, in high-speed rail systems and other means of transportation, in health systems, and in energy technologies where it would enable "an *energy superhighway* by supplanting copper electrical conductors with a ceramic superconducting alternative that has higher capacity while eliminating losses that typically occur during transmission".[32]

More powerful computers and more fundamental advances in computational methods, taking advantage of new (superconducting) materials, according to Kaku (see endnote 2), would lead one to assume that in the future "everything would have a tiny chip in it, making it *intelligent*"; we would then be living in "a world populated by robots that have humanlike characteristics!" (see endnote 2). Likely, then, technology will drive ethics and not the other way around.

Newness in future computing and in computers themselves[33] would allow *abundant avenues to knowledge and its use and misuse.* Through the Internet, developing nations taking advantage of the information revolution, will be able "to take a shortcut to the future", to build on existing intellectual capital. As we have noted earlier, we shall all be changed whether by ubiquitous computing or by virtual reality. More powerful computers and more fundamental advances in computational methods will undoubtedly take advantage of new materials, including superconducting materials.

4.2.4 Energy (New Sources, New Carriers, New Transformations of Energy, and New Ethical Issues)

Frontier science-based energy technologies promise abundant, "clean" energy, intelligently conditioned to meet the needs of modern technology; safer electrical energy from nuclear fission and abundant clean energy from controlled nuclear fusion; more efficient, cheaper and larger-scale renewable energy sources (such as solar and wind) with energy storage and fuels capabilities; transmission of copious amounts of electrical energy over long distances with low losses[34] (see Chap. 7).

Energy is and will continue to be critical for society An incessant flow of energy is the basis of modern civilization and indeed of all life. Technology may be limited by not just the amount of available energy for its use, but also by the forms of energy that are available to it. For instance, technology today (information technology in particular) is dependent on the availability of energy in especially conditioned forms. Generally, new ways to access known forms of energy and new sources of energy will be sought, and new energy transformations and energy carriers will be searched for. What will succeed electricity as an energy carrier? Would photons replace electrons as energy carriers?

Energy is the key to achieving stability of the Earth's climate Energy production and use will continue to raise fundamental challenges and serious concerns regarding their adverse impact on the environment and climate change. The energy-climate era will thus continue unabated. Hence, up and until humanity obtains abundant "clean" energy, it needs to slow-down the use of "unclean" energy and to reduce its consumption of energy through energy conservation and improvements in energy efficiency.

Energy raises moral issues as major factor of social wellbeing Ethical questions are raised about the use of energy and the access to energy. World poverty is basically energy poverty; to eradicate poverty we must satisfy the basic energy needs of poor people. Countries where a large fraction of their population lives on miniscule incomes of less than $2 per day, have little or no access to electricity.[35] Developed countries consume up to a thousand times more electricity per person per year than the underdeveloped. There is in fact a clear relationship between the consumption of electricity and the GDP of a country (see Fig. 2.8 in Chap. 2). The high-energy consumption by the developed countries today affords their citizens the greatest choice in human history; lack of energy thus means absence of choice. The future is, therefore, clear: *Escape poverty through provision of energy and particularly electricity; access to affordable energy may be regarded a fundamental human right and a moral obligation of civilization* (see endnote 34).[36]

In the fight against world hunger, energy will continue to be the key factor. Humanity will thus continue its efforts to make its use of energy compatible with human survival, need and dignity, and with its obligations to the Earth.

4.2.4.1 Future Energy Sources and Needs

As to the primary energy sources now available (fossil fuels, renewable energy sources, and nuclear energy from fission) a realistic assessment is needed of the potential of each taking into consideration their advantages and disadvantages, their direct and indirect costs, and the dangers associated with the technology of each kind of energy source. The challenges are and will continue to be many (see also Chap. 7):

- *For fossil fuels*: better and cleaner combustion, lower CO_2 emissions, capture of CO_2, replacement of oil with new less polluting fuels, energy conservation, increase in energy efficiency and gradual reduction in the use of fossil fuels.
- *For renewable energy sources*: for these mostly dispersed, intermittent, and cleaner compared to fossil fuels energy sources, is needed expansion,[37] storage capabilities for renewable energy on a large scale, transport of renewable electricity over long distances with small energy losses, development of renewable energy sources in the rural areas of the poor regions of the Earth, and cost reduction. The significance of new materials to meet these needs cannot be overemphasized; for instance, new materials for achieving lower costs and higher efficiencies in energy storage and transport of renewable electricity.
- *For nuclear power*: the major advantages of nuclear power are energy security, reliability, and reduction of greenhouse gas emissions compared to fossil fuels. The major disadvantages remain the high initial costs of nuclear power plants, the need of high-level security and safety of nuclear power plants,[38] strict international inspection mechanisms, prevention of proliferation of nuclear materials, and better management of spent nuclear fuel and nuclear waste (see endnote 18).

Despite the disadvantages, and the massive turn to solar energy and other renewable energy sources for electricity generation, it seems that humanity will expand the generation of nuclear power from nuclear fission in the future. For this, the issues just mentioned need to be addressed. Today, there are in use 440 nuclear power plants for electricity generation, which produce about 16% of the world electricity and more than 30% of the electricity in the EU (in some countries such as France, nuclear electricity exceeds 70%). Recently, renewed interest has been expressed for higher safety, lower amounts of nuclear waste, small size (100–400 MW) nuclear reactors, and several new reactors are being built by a number of countries such as India, China and Russia (see Chap. 7). According to Wang,[39] almost half of the nuclear reactors under construction world-wide are in China. China needs nuclear power to meet its energy demands and carbon-free targets, but it also "needs to do more to reform and strengthen its nuclear-safety system to match its expansion, including greater transparency" (see endnote 39). (See further discussion on nuclear power, breeder reactors, and fusion in Chap. 7).

Undoubtedly, in the future, humanity, through science and technology, will discover new useful sources of energy, new useful forms of energy, new technologies to access existing forms of energy and new technologies for more efficient use of energy. It might become possible to grow biomass using seawater[40] or to develop artificial photosynthesis. Ways might also be found to produce fuels by chemical recycling of CO_2 from natural, industrial, or other sources, or by capturing CO_2 from the atmosphere. It might become possible to manage solar radiation in space before it reaches the surface of the Earth for generation of larger amounts of not-stochastically-varied renewable electricity (and/or return to space part of the solar radiation which otherwise would reach the Earth surface and overheat the planet).[41] New materials might permit more efficient conversion of solar photons to electrical energy, fuels, or useful heat, and thermoelectric materials might be employed to efficiently generate electricity using "waste" heat. Perhaps, as well, new forms of energy may find applications in the emerging technologies of quantum computers and quantum communications and lead to new faster technologies based on photons rather than on electrons. Electrons might be replaced by photons as carriers of energy and radically change energy production, transport and use. Perhaps, as well, more efficient technologies for the transport of electrical energy (e.g., using super-conductors) may make economically feasible the production of electrical energy in the deserts of Africa, Middle East, America and elsewhere and its transport to the consuming urban centers thousands of miles away, as many have dreamed.[42,43]

In the future, more so than in the past, breakthroughs will be sought in the way energy is used. Energy consumption must be tamed through conservation and efficiency and change in human behavior. Ways to produce goods and services with less energy will be pursued as, for instance, in agriculture where presently the industrial food system depends on fossil fuels in much the same way the electricity and transport systems do.

The role of energy will remain critical both for science and technology. The perennial flow of energy will remain the basis of modern civilization and of life itself. Energy conservation, the cheapest and cleanest form of energy, will demand more attention and clean, safe, secure and affordable energy will continue to be one of the greatest challenges facing us tomorrow. We need to continue our search for the discovery of new usable forms of energy.

How may humankind make its use of energy compatible with survival? We understand energy scientifically, we control it technologically, but we seem not to have yet mastered it as the major factor of social wellbeing. The moral dimension of energy will intensify in the future because humanity must make the use of energy compatible with human survival and human dignity. The general challenge to modern civilization and its values is clear: *Secure the necessary sources of energy for humanity today and for future generations and adhere to our obligations to humanity and the Earth.*

4.2.4.2 A Unique Form of Energy: Light

Light is a unique form of energy. It offers incredible ways of manipulating energy. The unique interactions of light with matter as photons of varying energy and as photon beams of varying intensity and duration (continuous or pulsed) by themselves or in combination with other kinds of radiation, for instance slow electrons, will be further explored, for the development of new light sources, pulsed-power optical switches, transformations of light into other usable forms of energy, fuels, and laser fusion (see Chap. 7). The significance of light and its crucial role in science and science-based technology will become more apparent in the future.

Beyond energy, the crucial role of light and the slow electron will continue to be at the basis of most future advances in science and technology and progressively in biology. Pulsed lasers and low-energy pulsed electron beams can be employed to modulate the impedance (insulation-conduction) properties of matter.[44]

4.2.4.3 The Concept of Energy and Its Philosophical Dimension

Elsewhere (see Reference[45] and Appendix), I have conjectured on the concept of energy and its philosophical dimension. In all likelihood, in the future we will see further developments regarding the concept of energy and its philosophical significance. The answer to the question "What is energy and where did it come from?" will, in all likelihood, be a subject of intense future study and reflection, as all sciences, and philosophy, will come closer to recognizing the profound role of energy.

4.2.4.4 The Energy-Climate Era

The energy-climate era will continue unabated, because energy is of fundamental significance for the stability of the planet. Measurements are needed of the main environmental parameters and better models to mathematically describe the Earth's climate. According to Kerr,[46] "the problem with climate sensitivity is that you can't just go out and directly measure it. Sooner or later a climate model must enter the picture and every model has its own sensitivity. As a result, climate scientists have long quoted the same vague range for sensitivity: A doubling of the greenhouse gas carbon dioxide, which is expected to occur this century, would eventually warm the world between 1.5 °C and 4.5 °C. This range – based on just two early climate models – first appeared in 1979 and has been quoted by every major climate assessment since". Kerr points out that modeling requires better understanding of clouds and aerosols – the biggest sources of uncertainty; also, more and better records of past climate changes and their drivers must be retrieved.

Clearly, the intensity of activity in this area will continue and predictions will improve. A key challenge for the future remains the transition to carbon-free energy.

References and Notes

1. This could permit the supplementation of laboratory experimentation by computation.
2. Michio Kaku, *Physics of the Future*, Anchor Books, New York 2012.
3. Douglas Mulhall, *Our Molecular Future: How Nanotechnology, Robotics, Genetics and Artificial Intelligence will Transform our World*, Prometheus Books, Amherst, New York 2002.
4. Freeman J. Dyson, *The Sun, the Genome, and the Internet*, The New York Public Library, Oxford University Press, New York 1999.
5. Donald Kennedy and Colin Norman, *What Don't We Know?* Science **309**, 1 July 200, pp. 75–102.
6. Michael S. Turner, Physics Today, September 2009, pp. 8–9.
7. Frank Wilczek, Physics Today, April 2016, pp. 32–39.
8. Francis Fukuyama, *Our Posthuman Future: Consequences of the Biotechnology Revolution*, Picador, New York 2002.
9. B. P. Abbot et al. Physical Review Letters **116**, 061102, 1–16 (2016); Coleman Miller, Nature **331**, 3 March 2016, pp. 40–42.
10. Λουκάς Γ. Χριστοφόρου, *Η Επαγωγική Μέθοδος της Φυσικής Επιστήμης (Από τα Μόρια στον Άνθρωπο;)*, Πρακτικά της Ακαδημίας Αθηνών, τ. 82 Α΄, Αθήνα 2007, σελ. 1–30); Loucas G. Christophorou, *The Inductive Method of Science (From Molecules to Human?),* Proceedings of the Academy of Athens, Vol. 82 A΄, Athens 2007, pp. 1–30.
    Based on: http://www.1sbu.ac.uk/water/molecule.html and http://ww.chem1.com/acad/sci/about water.html
11. https://en.wikipedia.org/wiki/Brain
12. Steven Rose, *The Future of the Brain: The Promise and Perils of Tomorrow's Neuroscience*, Oxford University Press, Inc., Oxford 2005.
13. Attributed to Mike Holderness, New Scientist, 16 June 2001, p. 45. See also, John D. Barrow, *The Constants of Nature*, Vintage Books, New York 2002, p.117.
14. Gabriel Popkin, APS News, April 2016, vol. 25, No. 4, p. 3.
15. See, for instance, Phillip Anderson, *More is Different*, Science **177**, 4 August 1972, p. 393; Paul Davies, *The Cosmic Blueprint,* Templeton Foundation Press, Pennsylvania 1988; Stuart A. Kauffman, *Reinventing the Sacred*, Basic Books, New York 2008; Stuart A. Kauffman, in M. M. Waldrop (Ed.), *Complexity*, Simon and Schuster, New York 1992, p. 122.
16. Joseph A. Tainter, *The Collapse of Complex Societies,* Cambridge University Press, Cambridge 1988.
17. H. Jonas, *The Imperative of Responsibility*, The University of Chicago Press, Chicago, Il. 1984; Loucas G. Christophorou, *Place of Science in a World of Values and Facts,* Kluwer Academic / Plenum Publishers, New York 2001.
18. EASAC Policy Report No. 23 – JRC Reference report, *Management of Spent Nuclear Fuel and its Waste*, July 2014 (www.easac.eu ; https://ec.europa.eu/

jrc); see, also, K. R. Rao, Current Science **81**, 25 December 2001, pp. 1534-1546.

19. New areas are emerging from science-based technology, such as nanomaterials, which come with new benefits and with new risks. See, for instance, *Impact of Engineered Nanomaterials on Health: Considerations for Benefit-Risk Assessment*, Joint EASAC— JRC Report, September 2011 (EASAC Policy Report No. 15).
20. A. F. Koenderink, A. Alù and A. Polman, Science **348**, 1 May 2015, pp. 516–521.
21. I. A. Walmsley, Science **348**,1 May 2015, pp. 525–530.
22. Quantum entanglement is a physical phenomenon in which the quantum states of two (or more) systems are described with reference to each other, although these systems are spatially separated. The quantum state of each system cannot be described independently and measurements performed on one system instantaneously influence the other entangled system(s).
23. *The probabilistic nature* of quantum theory allows atoms and other quantum objects to store information that is not restricted to only the binary 0 or 1 of information theory but can also be 0 and 1 at the same time endnote[6].
24. M. Raymer and K. Srinivasan, Physics Today, November 2012, pp. 32–37.
25. Liquid nitrogen can be produced relatively cheaply.
26. https://en.wikipedia.org/wiki/High-temperature_superconductivity DOI:10.6084/m9.figshare.2075680.v2 (Pia Jensen Ray, Figure 2.4 in Master's thesis, Niels Bohr Institute, Faculty of Science, University of Copenhagen, Copenhagen, Denmark, November 2015).
27. https://commons.wikimedia.org/wiki/File:Timeline_of_Superconductivity_from_1900_to_2015.svg
28. Y. Li et al. J. Chem. Phys. **140**, 174712 (7 May 2014).
29. A. P. Drozdov, M. I. Eremets, I. A. Ksenofontov and S. I. Shylin, Nature 14964 (2015).
30. E. Cartlidge, Nature **524**, 20 August 2015, p. 277.
31. Sung Chang, Physics Today, July 2016, pp. 21–23.
32. See programs on High-T Superconductors by the US Department of Energy.
33. Explosive new developments lay ahead in many other areas such as *information and Internet technologies*. Light travels much faster than electrical signals, so using it to connect silicon chips would massively speed up calculations.
34. Loucas G. Christophorou, *Energy and Civilization*, Academy of Athens, Athens 2011 (ISBN: 978-960-404-216-6).
35. International Energy Agency (IEA), *Energy Poverty — How to Make Modern Energy Access Universal?* OECD/IEA, 2010; The World Bank 2011, *One Goal, Two Paths: Achieving Universal Access to Modern Energy in East Asia and the Pacific*, The World Bank, Washington, DC 2011 (ISBN: 978-0-8213-8837-2).
36. Loucas G. Christophorou, *Energy, Environment and Modern Civilization*, Proceedings of the International Conference on "*Transition to a New Society*",

Montenegrin Academy of Sciences and Arts, Podgoriga, Montenegro 2014, pp. 257–269.

37. The scale of renewable energy is still very small compared to that of fossil fuels.
38. A nuclear accident can be a real disaster (see Chap. 7; News & Analysis, *Radiation Risks Outlined by Bombs, Weapons Work, and Accidents,* Science **331**, 25 March 2011, pp. 1504–1505; Editorials, *Lessons of a Triple Disaster*, Nature **483**, 8 March 2012, p. 123; Issues & Events, *What Can Chernobyl Teach Us?* Physics Today, 1 April 2016, p. 24; James Mahaffey, *Atomic Accidents — A History of Nuclear Meltdowns and Disasters*, Pegasus Books, New York 2014).
39. Qiang Wang, *Nuclear Safety Lies in Greater Transparency,* Nature **494**, 28 February 2013, p. 494.
40. Ultimately, also, plants may be able to fix water from the desert air.
41. The Royal Society, *Geoengineering the Climate: Science, Governance and Uncertainty,* September 2009.
42. https://newint.org/features/2015/03/01/desertec-long/ https://en.wikipedia.org/wiki/Desertec
43. Robert F. Service, Science **309**, 22 July 2005, pp. 548–551.
44. Loucas G. Christophorou and James K. Olthoff, *Electron Interactions With Excited Atoms and Molecules*, Advances in Atomic and Molecular Physics, Vol. **44**, 2000, pp. 155–292; L. A. Pinnaduwage, L. G. Christophorou, and S. R. Hunter, *Optically Enhanced Electron Attachment and Its Possible Applications in Diffuse Discharge Switches*, Proceedings of the 6th International IEEE Pulsed Power Conference (P. J. Turchi and B. H. Bernstein, Eds.), IEEE Catalog No. 87CH 2522-1, 1988, pp. 81–84; L. G. Christophorou and S. R. Hunter, *Laser Activated Diffuse Discharge Switch*, Patent No 4,743,807, May 10, 1988.
45. Λουκάς Γ. Χριστοφόρου, *Ενέργεια: Επιστημονική, Φιλοσοφική και Θεολογική Διάσταση*, Πρακτικά της Ακαδημίας Αθηνών, τ. 89 A', σελ. 27–48, Αθήνα 2014. Loucas G. Christophorou, *Energy: Scientific, Philosophical and Theological Dimension,* Proceedings of the Academy of Athens, Vol. 89 A', pp. 27–48, Athens 2014.
46. Richard A. Kerr, Science **309**, 1 July 2005, pp. 75–102.

Chapter 5
Values of Society and Science

5.1 Introduction

We hold the view that in a world dominated by facts there is a place for values. A place for a system of common, shared values based on the demonstrated likeness among the values of the peoples of the world, and on the realization, that even in today's world of extreme diversity and individuality, the seemingly conflicting values of national and local cultures are but complementary. We are also convinced that by fostering a dialogue among civilizations, among national and local cultures, among the humanities, the religions and the sciences, and by looking at values from different perspectives, we will be able to better understand, identify, and safeguard humanity's common heritage of values and make it valuable.

While the subject is not new, the situation today is in many ways different than in years past, because enormously significant and complicated problems beset the world and threaten humanity: conflict; immigration and population increase; climate change; human condition and hunger; failing social, political, economic and ethical systems; the effect of science and science-based technology on all these and others. Increasingly, warnings are being heard that "our belongings to the human family have been shaken", that "our society is dehumanizing rapidly", that civilizations are "clashing" and that our values are "disappearing" or are "irreconcilable". It is argued[1,2,3] that the twentieth century has been marked by secularization of values and that future societies will be more rational and less humane as they become more scientifically and technologically advanced. The conflict between rationalism and irrationalism has become an intellectual and moral issue and affects values.

Powerful science-based technologies, especially in genetic engineering and biotechnology, strike at the roots of our understanding of human nature and some see in it the threat "that humanity may be debased in favor of some featureless representation of the post-human" (see endnotes 1, 2, 3).[4] The evolution of humanity is foreseeing by others[5] to "reach a singularity point in time where it will simply disappear to the benefit of some new beings." A quote from one of the scientists who

L. G. Christophorou, *Emerging Dynamics: Science, Energy, Society and Values*,
https://doi.org/10.1007/978-3-319-90713-0_5

contributed to *The Millennium Project on Future Global Ethical Issues (Forecasts of Value Changes)* written for the United Nations (see endnote 5) speaks clearly to the issue; it reads:

> "I firmly believe that ethical considerations based on tradition and religious beliefs will tend to disappear and give way to a more scientific, technological and economical world; a world in which the human being, the individual, and the traditional concepts of ethics will tend to disappear to give way to a new ethics of pragmatism, technology and collectivism. The traditional nucleus of society – the family – will disappear; the concept of offspring will disappear; the human being will be seen by itself as a couple of chemical reactions inside a bag. Birth and death will not be the basic points of life but singularities of machines. The machine society in which the human being is just another machine, that is the ethics of the future; no ethics at all as we see it today; no values at all as we see them today. Good and bad will have no meaning for the future generations."

An utterly pessimistic view of humanity's future!

Today, everyone speaks of values, but often understands their meaning differently. There seems to be "a chaos of values" as values become more subject-specific and more subjective. Modern relativism of values, many argue, may be a reaction to the ancient and medieval conception of values as absolute ethical imperatives, irreducible to reason and independent of man. While we can neither disregard what science says about human nature, nor we can regard ethics as beyond criticism of reason, it seems, as many have noted, that at the foundation of values lies an act of faith, which cannot be justified by reason alone.

Modern society's increased pluralism leads to more than one interpretation of values, and increased societal complexity affects values invariably. Like civilizations, values have no boundaries, but every nation (every culture) has its own. A new synthesis of values is needed that would make humanity's long-cherished dream of a common system of values come true.

The problems humanity faces today require common understanding and shared values. There is a need to resolve conflict, to avert the clash between peoples' values, to change our perception of resources and their consumption, and to address calmly and prudently the moral issues posed by the growth of science. We need a system of values that would "draw us together"[6] and would make us more sensitive to human injustice globally. As Huston Smith put it, "before peace and democracy we need to declare global war against poverty" (see endnote 6), because poverty deprives people of hope and people need hope to live in peace with each other.

In this Chapter, we focus on human values: their origins; their universality, commonality and complementarity; their significance for humanity; and how values are affected by science and affect science.

5.2 The Precepts and Concepts of Values

Traditional values take their meaning from precepts and concepts, which are sources of particular rights and responsibilities.[7] The precepts and concepts of values are profound; they join human beings into societies and concomitantly preserve their individual freedom. Where do precepts and concepts come from? From man's microcultures through dialogue; from the ten commandments, the golden rule, the sermon on the mountain, that is, for the most part through religion; from the "wisdom of the elders" of various cultures and from the poets and the philosophers of years past, like those of ancient Greece. The precept is a command: "thou shalt not steal", "thou shalt not lie"; only if those precepts are assumed valid will honesty be a value (see endnote 7). Values, thus, take their meaning and stability from precepts and concepts which are claimed to be unchangeable because they are either "revealed", or inspired, or self-evident. Such values (e.g., truth, justice, freedom, respect, dignity, love) are postulational and primary in their origins and they invite commitment; they are meaningless in isolation (see endnote 7). *It is the agreement of people upon a precept or a concept that makes the associated value valuable.*

Values are embedded in other values. Their value is implicit; they owe their value to the existence of other values.[8] For instance, each of the values of justice, freedom and dignity is qualified by the rest; the extent to which we respect a person's right to life is qualified by our sense for justice toward that person; the extent to which we respect individual rights depends on the degree to which we value individual freedom; the extent to which we respect our fellowman's dignity qualifies our tolerance of his ways of life and our obligations to his basic rights. This embeddedness of values allows for their mutual interaction and feedback, their mutual accommodation and indebtedness; it demonstrates the existence of a fundamental unity among human values on the basis of which humanity can converge onto a universally acceptable value-judgment system. A convergence based on *unity,* not *identity,* of values and a commonality of *individual and collective conscience,* itself based on the capacity of man to acknowledge norms and determine his conduct accordingly. In such a system, as will be discussed later in this Chapter, we would identify *universal* and *common* human values and we will treat other acceptable human values as *complementary* to those two "traditional" values.

5.3 Traditional Human Values

5.3.1 Values of Faiths and Cultures

5.3.1.1 Values of Faiths

Throughout history and in all civilizations, we encounter the transcendental values of religious faith from which, for millennia, billions of people have derived ethical standards. The main religions, especially Christianity, are anthropocentric. In the Christian belief, God has drawn out of the evolving cosmos a world of persons,[9] made *in His own image*. The essence of man, then, according to Christianity, is defined by the image of God in him.[10] In the depths of man's religious faiths lay the values of reciprocity, love (αγάπη), respect for life and human dignity, justice and truth, recognition of the common origin of all people, and the commitment and dedication to great precepts which may vary from religion to religion but point, just the same, to the truth common to every religion. Fundamental, thus, is the belief of many religious faiths that God is the supreme source of values, that religious values are a "revelation" of God to man and are therefore transcendental.

Distinct among the religious values is that of a "person". According to Christianity, personhood belongs to every human being by virtue of its unique relation to God who created it in His own image (see endnote 10). There have been, however, great variations and changes in our conception of what it is to be a person and some of these are connected with the impact of science. For instance, in the view of many biologists, man is just a collection of molecules, "an accident on the stage of evolution", fundamentally no different from any other living creature. Others maintain that there is no such thing as "human nature" or that altering it is not ethically problematic. Still others argue that man's own values grew out of his evolutionary origins and his struggle for survival[11,12] and are thus relative and never absolute. A growing consensus upholds the view that there is a genetic heritage of value and thus one can never properly analyse a system of human values independently of the information stored in our genomes.[13] Most researchers agree that genes influence human behavior. Others, like Gregory Stock,[14] argue that biological enhancement "will eventually challenge our basic ideas about what it means to be human" and "progressive self-transformation could change our descendants into something sufficiently different from our present selves to not be human in the sense we use the term now." Developments in genetic engineering strike at the root of our understanding of human nature and engender, in the view of many, the possibility that humanity as a value may be "debased in favor of some featureless representation of the post-human".[15,16] Such notions raise fears "that we are in the process of redefining ourselves as biological, rather than as cultural and moral beings" (see endnote 16), that "man's final conquest has proved to be the abolition of man".[17]

Many philosophers consider a person to be a moral agent, not just a cognitive or a rational agent, and hence the view of man as no different from other animals represents radical depersonalization of the human being.

While religious values are viewed as permanent and universally applicable, there clearly is particularization and differentiation amongst them that accounts for the diverse ways in which religious values adjust to human condition, the peculiarities of cultures and new knowledge, especially the knowledge provided by science. For instance, ethics and duties are traditionally grounded on reciprocity: "Thou shalt love thy neighbor as thyself". But today one may ask "who is my neighbor?"; the seven or so billion people on Earth? Clearly, one's neighbors far exceed those in years past and the meaning of "reciprocity" has been enlarged to include one's duties to his distant neighbor and to his future neighbor. Science has imposed on us the moral duty to protect the dignity of man everywhere, today and in the future. Modern science has stripped us of the right to isolation. It is, however, through religious values, not through science, that we can see tribes, races and cultures – people however remote – as human persons, worthy of respect having the same basic rights as ourselves.[18]

Let us carry this discussion on science and religion a little further focusing on their similarities and their differences, and on their possible mutual accommodation.[19] Science is supremely inductive while religion is supremely deductive; the former relies principally on the properties of the parts to extrapolate to the properties of the whole, while the latter goes the other way. Physical facts are never subject to faith. They can be ascertained by the method of science. The relationship between faith and reason (religion and science) changed throughout history and varied between cultures and civilizations.

Neither Jacques Monod's statement "that everything is accidental", nor Albert Einstein's denial of chance because God knows "how the dice will fall" are facts of science; they are private opinions. Science cannot be used to disprove the existence of God, just as religion cannot be used to disprove the validity of the physical law. Science must shy away from human values, moral matters and the knowledge that is beyond its domain and should resist any attempt to be dragged into scientism.[20] Although science changes man's perception of himself and his relation to the universe, what is good and what is evil cannot be judged by the standards of science. "Science deals with things not people" Marie Curie once said.[21] Conversely, religion must accept the proven facts of science and the scientific description of the physical world, preserving its independence and autonomy, free from the bonds and restrictions of an evolving science. When religion intruded into science[22] the consequences were disastrous. They may be worse when science intrudes into religion as it has been attempted lately by some biologists[23,24,25,26,27,28,29,30,31] and cosmologists.[32]

Neither science nor religion can be eliminated; they are parts of human existence and experience; they are here to stay – man's need to know cannot be satisfied by either alone. Science and religion speak in different languages, but they both employ paradigms to convey the complementary aspects of reality and truth (see endnote 19).[33] They both speak with certainty and yet in both science and religion there exist unanswered questions and plenty of doubt. There are limits to perception in religion as there are limits to the power of the "scientific method". Neither does the scientist live his personal life by scientific concepts alone, nor does he need advice from religion to conduct his research. There is a real and beautiful world out there independent of and beyond science. Indeed, man is neither what science says he is, nor what religion says he is. He is what both say he is, and a great deal more. Beyond divisiveness and ideological divides there exists a common humanity comprised of scientists, believers, nonbelievers and everyone else. There is thus a need for continuous dialogue between science and religion and mutual accommodation that recognizes their separate ("non-overlapping magisterial"[34]), but complementary realms.

Let us carry this discussion on the similarities and the differences between science and religion/faith a little further still. Personal knowledge and understanding through faith are naturally affected and conditioned by other people, although deep down they remain utterly private. This is profoundly different from the knowledge and understanding of the physical world through science, although even scientific knowledge and scientific understanding has a personal and intuitive twist. The answer(s) to questions which fall within the domain of science are public and restricted/specific, the answer(s) to questions which fall within the domain of religion/faith are private, unrestricted conceptually, and different from one person to another; there is one science, but there are many religious faiths.

In science, when new findings cannot be rationalized, scientists look for consistency both among the new scientific findings themselves and between the new findings and the uncontested facts of the rest of science. When consistency evades, and reference to existing knowledge leaves unexplained the new findings, what do scientists do? They allow themselves the freedom to introduce new constructs and new paradigms with which they explain the otherwise "irrational" new scientific findings. This, for instance, was what led Max Planck to introduce the concept of quantization of energy. And this is what led to the introduction of the concept of complementarity into physics by Niels Bohr and the construct of particle-wave duality to rationalize unexplained scientific data: the same physical entity is described by two different constructs, both of which complement scientific understanding of one and the same reality, which would have been impossible to fully characterize with only either form of description. Reality, thus, needs to be looked at from different angles for a fuller description and comprehension, and this may be true for both science and religious faith. Like the physicists, believers and mystics face the paradoxical experience of reality and attempt to "know" it via contradictory but complementary approaches, and at times also through a common "understanding" of the lack of understanding. There are situations where "knowledge" is shrouded in doubt to such a degree, that "knowledge" is expressed antinomically, by negation – invisible, incomprehensible, unknowable God – an admission of the uncertainty and the limits of religious knowing.

The means science uses to explore nature (scientific instruments, computers, theories) differ from those of religion (prayer, ritual, worship, music, symbolism). Each set of tools – those of reason and those of faith – provides complementary knowledge for understanding the common world. There are things that can be accessed only by reason and there are things that can be accessed only by faith, although there always seems to be an intuitive element in science just as there is a rational element in religious faith. Generosity and integrity are basic elements in the method of science and in the way of religion from which grows a sense of dignity that links the values of science and the values of religion. Such *complementary aspects* induce mutual readjustment and help moderate the differences.

The big and complicated questions humanity faces today bring science and religion closer to each other, although on the surface they seem to be doing just the opposite. Unquestionably, there is today tension between science and religion, which in some respects differs from that in years past. Today, the tension concerns both the consequences of the new scientific knowledge on the teachings of religion and the consequences of science-based technology on humanity, while in the past the tension tilted toward the consequences of basic new scientific discoveries on the teachings of religion. The tension between science and religion is but a symptom of the continuous struggle between the two for mutual accommodation.

The meaning of universality in science differs from that in religion. Universality in science means that factual statements and theories – basic scientific knowledge – are common human heritage everywhere and at all times, independent of the culture, religion, and nationality of the scientists. Universality in religion entails that in the entire history of humankind certain kinds of feelings and considerations of transcendence, although differently expressed, can be traced throughout history. Today, there is a great deal of misunderstanding between science and religious faith largely due to ignorance of the other side by the "disciples" of both. Many have written on the subject, but no one seems to have dealt in depth with two fundamental qualities of enormous significance shared by both science and religion, namely *universality and complementarity.* Appreciation of these two characteristics requires knowledge and understanding of both science and religion and recognition that neither can get rid of the other, nor, if that were possible, it would be desirable for society.

Though science is ontologically non-committal, it is not ethically neutral. Through its epistemology science provides a rational basis for ethical decisions by clearly defining the facts. It induces the adjustment of religion to the condition of modern man and the evolution of values. The values and ethics of modern man cannot be based on science, but they cannot be divorced from science. Not everything in life is accessed by reason and science; faith remains for billions of people a complementary source of knowledge, but "only where faith is can science exist with faith" (see endnote 19).

5.3.1.2 Values of Cultures

In the springs of national cultures lay the micro-cultural value-judgment systems of the peoples of the world. Indeed, history teaches that no culture is possible without agreement on a foundation of common values of its own to guide the behavior of its people and the actions of its governance. Culture-specific values reign supremely.[35] They qualify, articulate and affirm the cultural heritages of peoples.

At the *International Symposium on Universal Values,*[36] organized by the Academy of Athens in connection with the 2004 Olympics in Athens, the late philosopher Constantinos Despotopoulos[37] nicely described the philosophical foundation of the values rooted in the cultural heritage of ancient Greece. There, in ancient Greece, man is the supreme value, the measure of all things; his attributes are wisdom, virtue, moderation, balance, civility, responsibility, duty, patience, heroism, greatness; he respects life, nature and the law, and knows himself and the limits of his freedom. At the very foundation of Greek values lies Protagoras' dictum that *man is the measure of all things* and Aristotle's thesis that *man is the supreme value par excellence.*[38] In such a philosophical perspective, values transcend history and are in their essence universal. Other cultures have had and are having different outlooks. There has been, however, an underlying unity in the micro-cultural value-judgment systems of man.

We have repeatedly stated in this book that man is unique not because of his physical or biological composition, but, because, as far as we know, he is the only bearer of thought in the universe,[39] he is a cognitive, rational, and moral agent endowed with freedom and responsibility. Man, not science and technology or the rest of life, is accountable for the choices that either enrich or diminish the quality of life. He alone has the ability to access truth and to uphold human values. Will future scientific technology make him freer or must he guard against science and technology if he is to remain a free human being?

Recent scientific advances are seriously impacting traditional human values, whether universal or micro-cultural. At times, it might even be necessary for values to adjust to certain needs of scientific research, as for instance, in the case of stem cells.[40] Nascent life, it is argued (see endnote 16), should neither be destroyed for the sake of research nor be looked at as a natural resource to be mined and exploited. Questions raised by stem cell research are but the forerunner of questions we will be confronted with in the future; biotechnology and genomics offer power to alter and to control the phenomena of life – in plants, in animals, and increasingly in humans; genetic engineering strikes at the root of our understanding of human nature, and some see it as a threat to humanity as a value; genetic information on individuals would have profound implications for human freedom and dignity; human embryonics would make possible control over human heredity and reproduction; genetic and reproductive technologies, claim others, would have the potential to change what it means to be human. A part of society is thus worrying that the more man's scientific knowledge ascends, the more man seems to be devalued by it. Independently, traditional human values will be challenged and new rules, standards, and professional ethics will be called for as science marches on. In the new field of bioethics,[41] as in many other similar situations, "the evils we face are intertwined with the good we seek – the supreme values of modern life": cures for disease, relief from suffering, preservation and prolongation of life and so on.

A common universal value-judgment system transcends the value-judgment systems of traditional cultures, but it does not replace them. Many of the cultural values are in essence complementary, other expressions of the universal values and alternative ways of perceiving their richness and can, thus, coexist. It is however argued[42] that "until very recently virtually the entire human history is expressible in terms of cultural divergence and that the cultural specificity of values makes it impossible for values to become universal".

5.3.2 Contextualization of Values

Values arise in the lives of persons and are thus influenced by the circumstantial conditions of life; they are related to the values of the local environment which gives them context and constrains their meaning. Values, therefore, are in practice *contextual*. The values that existed in ancient Sparta and in ancient Athens, the values that existed in Europe in the Middle Ages, the values that existed in the plantation life of the American South, the values that existed in societies occupied by colonial powers, the values of peoples under fascist dictatorships, under the totalitarian regimes of communism, or under the extreme fanatic state-rule of late, are not identical with the values of peoples in twentieth century industrial "democratic" societies. There have been societies in which stealing from or lying to the enemy was permitted, history reminds us. In such circumstances, then, honesty is not an unqualified value, but it is determined by an appeal to a higher precept, a superior value, and the value context within which the act is embedded.

This *contextualization of values* is clearly seen in situations where abrupt changes in peoples' lives took place as a result of changes which ended totalitarian rule. It is instructive to refer to a recount given in a paper[43] by Professor Jüri Engelbrecht of the abrupt changes in values in Central and Eastern Europe (CEE) in the twentieth century that occurred because of changes in society caused by two World Wars. Freedom, he says, was for the people in CEE a basic value, but he then asks "freedom from what?" He goes on and compares the values of the CEE peoples under a century-long totalitarian rule and recently in the European Union this way: The totalitarian system that governed most of the CFE countries, "was spiritually very oppressive and brutal", constantly pressing the moral and value systems of these peoples. The escape from authoritarian rule meant freedom, personal independence and rational choice, a change from survival values towards individuality and diversity, towards the right to have different values and to honor them. Sadly, after all that trying experience under totalitarian rule, he sees in the lives of these same people, now in the EU, "freedom running away from responsibility" and diversity of values eroding. In practice, then, values are displayed differently in the behavior of people, and the values – and their value (αξία) – a society upholds evolve depending on culture and context. Of course, even in those regimes, people might have secretly held totally different 'hidden-from-the-state' values that sustained them in their private life and resistance to the system.

Relevant, I believe, to the contextualization of values is also the question: How are values changed or distorted in a society and how long does it take to change or to distort a given society's values? In contemplating such questions, I recall a discussion my wife and I were part of during dinner at a colleague's house in Kaiserslautern in 1991, where a number of guests from former West and East Germany were present. Following a most engaging discussion of the conditions that prevailed in former East Germany before and after the fall of the Berlin Wall, and a fascinating account of what a teacher in former East Germany was teaching the German youth before and after the changes, I turned to her (the teacher in question) and said: "How devastating an experience this must have been for your young students to be taught a set of values and ideals one day and to come the next day and be taught another, exactly opposite, and be told that the 'old stuff' was 'totally wrong' and should be let go." To which the teacher calmly but firmly replied: "Do not worry, they never believed what they were told before anyway!" I understand that this is probably because the city at which the teacher was teaching is located within the range of the then West Berlin TV stations! No secrets, no false values, the students knew the facts! The means provided by science and science-based technology penetrated the iron curtain!

Indeed, this is not unlike statements made to me by top scientists, in European countries then under communist rule, for instance in Poland and Hungary. In Poland, a colleague told me "we are 90% communists, but 95% Catholics" and in Hungary another colleague told me that the then regime made the mistake to leave the children in the hands of their grandmothers and they "taught them the right stuff." Similarly, others described themselves as "an apple, red outside but white inside".

It is thus an error to suppose that the norms of a society can be changed by imposing new principles and rules of conduct. Beyond contextual meaning and distortion of values, time-and-again, the value of a given value and its essence remain fundamentally unchanged and this is basically the value's value, to be there whenever a person needs to turn to it.

5.4 Universal, Common and Complementary Values

We distinguish three groups of values: *universal, common* and *complementary*. The first are primary, the second can be primary or culture-specific-not-in-opposition with the primary, and the third can be other values that are consistent with the former two and not in contradiction with them; the third group can include values related to new areas of human experience such as the values of science, human readjustment to nature, energy, climate change and the environment. We shall identify such values and attempt to qualify their content and value.

5.4.1 *Universal Values*

The Greek philosophers Plato and Aristotle (teacher and student) differed in their description of values as universal or as cultural. Plato argued that morals are based on the knowledge of universal ideals and therefore have a universal character, while Aristotle maintained that ethical rules should always be seen in the light of the traditions and the accepted norms of the community. In the course of history both views have been adopted.

History tells us that there is a heritage of values embedded in human cultures, national civilizations and religions, that constitutes the foundation of man's intellectual and moral tradition and the standards by which we judge the significance of life, and in the light of which justice and injustice, freedom and slavery, good and evil are in sharp contrast. These values determine our virtue and honesty, our friendship and honor, our tenderness and goodness, our dignity, our love for each other, and they draw us together. They are not generated by science and they are not negated by science, but they are affected by science. They have changed through time and place, but remained in their essence timeless, ***universal.*** They are freely adopted by people and can accommodate cultural diversity by preserving cultural individuality — are *universal in their diversity*. Universal values are thus common values based on the proven similarity among the micro-cultural value-judgment systems; values the world agreed upon and freely adopted. They bear no similarity to the agreement and acceptance of scientific knowledge as universal common value. The substratum of universal values includes: *respect for human life, freedom and justice; commitment to peace, human dignity, reciprocity and love (αγάπη).*

Universal values are prerequisites for universal ethics. Such ethics is not nationally, culturally, or religiously bound; it is needed to order human actions and to regulate human behaviour. Universalism in ethics presupposes existence of shared values and agreement on moral norms. Total universalism in all ethical questions is probably neither possible nor desirable.

While at the foundation of man's ethics is not the knowledge of the scientist or the technical expert but rather the knowledge of people of good will, man's ethics needs adjustment of the human relationship to Nature. The way we face this challenge is related to the way we see the human person: as a being separate from Nature or as one species among others embedded in the intricate web of natural processes that embraces all forms of life. Clearly, not everything in Nature is of the same value! I recall the statement made by a distinguished research professor at a major US university, who was conducting pioneering experiments on radiation-induced cancer using mice, when he got the news while at a scientific conference that his laboratory had been ransacked by people advocating "animal rights": "I rather kill a mouse to save a human life", he simply said.

5.4.2 Common Values

There exists a common humanity beyond all divisiveness. How, then, can the world's multiple cultures mutually converge to a common frame of reference for value judgment that would help us bind our fragmented world and address the common problems of humanity today, be them environmental, biomedical, nuclear, social, political, or problems of injustice, or war and peace, or, further still, problems pertaining to the image of man and the excesses of his own power? How can we build a common understanding, shared values, and mutual trust among religion, humanism, and science for the benefit of all in a common hopeful future? Such commonality cannot be based on utility alone. Each common value must integrate diversity in its own content; each must unify humanity with its differences. A common value-judgment-system transcends traditional cultural values but does not replace them.

Establishing a stratum of common values has been a cherished dream of humanity for a long time. Today, many argue that what is new is the urgency of this need. Yet, the process can neither be rushed, nor be mandated. It must grow from the impoverished roots of the human past and the determination of the present humanity. A system of values common to humanity should follow the complementary way: *accept all universal values as common and identify other common values which are not universal but have common complementary unifying cultural elements*. Commonality should, furthermore, be sought between the values of science and those of society for beyond the scientific fact and its potential use lies the common system of societal values on the basis of which decisions are ultimately been made.

Earlier in this section we stated that common human values can be universal or culture-specific-not-in-opposition with them. The unity of these values *is* prerequisite for their commonality. The substratum of common values includes the universal values themselves (freedom, justice, respect for human life, reciprocity, etc.) and also values such generosity, caring, fairness, truthfulness, trust, tolerance, and commitment to human-dignity-based values including the material basis that this obligation entails; values to safeguard free inquiry, free thought, free speech, and commitment to nonviolence. These common values are central to human rights[44] and freedoms.

A major hindrance to the emergence of common values is prejudice, fear, and a loss of faith in human institutions including those of science and religion. Many see the need for moral principles to live by, but they seriously doubt that science can handle such a role as supreme judge and master of the entire society. They fear scientific control of society and reject scientific materialism and reductionism as the true determinant of human life. Science by providing the means for communication among cultures aids the emergence of common values.

5.4.3 Complementary Values

Besides the universal and common values we have just referred to, there can be partial or total convergence on a number of other values some of which are new and emerging and address new problems, issues, facts, or human condition. These values are not necessarily universal or universally common; we treat them as complementary to the values of the other two categories as long as they are not in conflict with them. Four important areas are mentioned below which could help exemplify the need for complementary values.

(i) *The values of science*: Modern man is morally burdened to make the values of science as much a part of our lives as are the universal and the common human values; we act implicitly as scientists and as human beings. Just as there is implicitness among the universal human values, just as there is implicitness among the common human values, just as there is implicitness among the values of science themselves (see endnote 19),[45] there should also be implicitness among the values of science and the universal and common human values. The values of science are rooted in the practice of science and they can thus be termed "instrumental"; they qualify science's functions, but they are not substitutes for the universal and the common human values.

(ii) *The values and ethics of the environment, climate change, and energy*. Old and new values alike are needed to raise the conscience of humanity to protect the Earth and the life it supports. Science has decidedly helped change man's ethics and behavior in this regard and has stretched modern man's ethics both in time and in distance ("delocalized" ethics[46]). Coexisting with traditional ethics, "delocalized" ethics imposes on man new responsibilities. The same holds true with many of the issues we discussed in Chaps. 2 and 7 relating to energy, its impact on the environment and its role in eradicating human poverty.

(iii) *The values and ethics pertaining to the profound impact of science on society and on man himself*. New realities have been brought about by science and science-based technology in this area and are referred to throughout this book, for instance those associated with the genome and the internet.

(iv) *The global civic ethic and "consensus" values.* Global ethics normally comprises a common moral minimum of values shared by "all" cultures and religious traditions and a set of rights and responsibilities constituting a "civic code" based on those values. It may, in fact, be necessary to establish "consensus values" and "consensus codes of conduct" closely associated with culturally-induced beliefs of individuals and groups. There is, in fact, a proliferation of meanings and definitions of values of all sorts, some of which are identical with practices and professional ethics and clearly cannot stand by themselves but by reference to the universal/common values. In a democratic society, there's bound to be many conflicts over these, and science and technology will certainly be caught up in many of them.

5.5 Values of and in Science

Science deals with the physical world and with questions that can be defined scientifically, can be studied scientifically, and have a chance to be answered scientifically. Science unravels the beauty of the physical world like no other human activity; indeed, the basic knowledge science provides can be considered a value in and of itself. However, as was mentioned elsewhere in this book, science is not the only way to the truth. Beyond science, beyond the physical and the biological, beyond that which can be proved by the method of science and can be measured by the scientific instruments, lay the spiritual, the cultural, the intellectual traditions and values of man, and the teleological concepts of philosophy and religion of which science does not speak. Science does not discover values; its standards cannot judge what is good and what is evil; science deals neither with ethical judgments nor with the ultimate meaning of life. I know of no physical law which demands respect for human rights, or the love of my neighbor.

While science *per se* does not deal with human values, science is not free of values in the execution of scientific research and in the application of scientific knowledge; science and science-based technology impact values and can serve as a means to values. There is a fundamental role of values in science, just as there is a fundamental role of science in values.

There are values ***in*** science and there are values ***of*** science.[47,48]

5.5.1 Values in Science

It has been correctly stated that one can only work in science if he/she values truth. The search for truth in science imposes on the researcher a moral conduct, which is not unlike the moral conduct of a person in the broader society. Science confronts the work of a scientist with the work of his colleagues and cannot survive without justice, honor, and respect amongst them. Science, furthermore, is based on free communication among the scientists and on mutual trust. Freedom of thought and speech, justice, self-respect, integrity, generosity, and tolerance of differing views, are all values recognized in the past – long before modern science – as necessary for the survival and the civic functioning of society. On those very values relies science for its functioning and on them rests the freedom and the responsibility of the scientist. Thus, while the scientific picture of the natural world is constantly changing, the values on which science and scientific behavior depend remain in essence the same, timeless, universal values.

Besides the ethics which is applicable in the conduct of research, a scientist has a duty to inform society about his research findings and is accountable for the choices he makes that either enrich or diminish the quality of life. Inescapably, however, as long as science is a human activity carried out by individual men and women it must at bottom line remain subjective, prone to error, human failings and occasionally sheer fraud. The scientific ethic is inescapably conditioned and invariably contextualized by location, imposition of one's employer or sponsor, or the mission of the research. The scientist's struggle to reconcile all these is often a challenging task.

5.5.2 Values of Science

The values of science do not derive from the virtues of its members. They have grown out of the practice of science because they are the inescapable conditions for its practice.

There are values *of* science, which characterize its functioning: *originality, rationality, objectivity, verification of knowledge, discovery and correction of error, respect and acceptance of the proven fact, unification and coherence of scientific knowledge, cooperation, universal participation in science, humanism.* Humanism is a multi-dimensional value of science for, as we have stated earlier,[49] "if deep in the essence of civilization lies the emancipation of humanity, society cannot be truly civilized without science". The values of science are not substitutes for traditional human values, as some have advocated, but can be accepted as complementary.

Broader acceptance of the scientific values by society rests principally with the conduct of the scientist himself, on how he adheres to the scientific values and in practice shows their value. Today, this seems more difficult than in years past because the spectrum of people working in science, the dimensions of scientific research being done, and the nature of the places where it is conducted vary enormously. There seem to exist significant differences among countries in the conduct of responsible scientific research. Indeed, a number of international organizations, amongst them the InerAcademy Council (IAC)[50]and All European Academies (ALLEA),[51,52] have drawn attention to the need for "forging an international consensus on responsible conduct in the global research enterprise". The IAC report states that a "global research enterprise is emerging, which requires that the universal values of science be embodied in global standards of behavior that are understood and followed by all." Both the ALLEA and the IAC reports call for research integrity from the part of researchers, research institutions, public and private funding agencies, and scientific journals; they also call for proper allocation of credit of research results and eradication of misconduct in science. Both reports acknowledge that plagiarism is on the rise. These findings have led to calls for common training to ensure research integrity and shared scientific values and norms, for calls of "*codes of contact*", and for calls of reinforcement of ethical standards in science especially in the execution of research. Self-regulation of scientists has been suggested to address such problems rather than imposition of rules from above, but perhaps this is not sufficient in view of the diverse people doing science and the nature of the conflicts involved. The responsibility of the scientist can be at the individual or at the collective level and can be perceived differently by the society and the State. Two recent examples, mentioned in Chap. 3, which exemplify the sensitivity of society to what the scientists do are, first, the accusation and conviction of seven Italian scientists of criminal negligence and manslaughter for failing to adequately warn residents before an earthquake struck the city of L'Aquila in central Italy in 2009 killing more than 300 people[53,54,55,56] and, second, the adverse reaction to allegations of manipulation by scientists of environmental data and their analysis.[57]

It is difficult to identify incontestable ethical constraints so fundamental that they could have a universally imperative character. Drenth[58,59] suggests research is not justifiable if (i) before, during, or after an experiment or the gathering of empirical data, *unacceptable* damage is inflicted upon the object of the research (whether this concerns people, animals, Nature or culture), or upon the wider social or physical environment; (ii) the nature and/or consequences of the research are in conflict with basic *human values*; (iii) it contravenes *solidarity*, firstly with *humankind*, secondly with *posterity*.

A broad, continuous dialogue between science and society is therefore necessary, a dialogue based on wisdom and trust for if one issue is resolved another issue will surface. Unless the value systems of science and society reach mutual accommodation, neither the ability of man to resolve satisfactorily the issues raised by the advancement of science and the impact of science-based technology, nor the ability of society to optimize the benefit from the advancement of science can be accomplished. Both science and society will lose out.

5.6 The Impact of Science on Values

Science impacts values through both its content and its applications. The unity of the universe as is uncovered by science, and the content of science, change man's understanding of the cosmos and his place in it. Similarly, science and science-based technology contribute to the evolution of moral consciousness and ethics by setting before man choices of increasing complexity and by providing much needed input into making ethical decisions in practical circumstances.

Science induces the adjustment of religion and ethics to the condition of modern man and the evolution of values, for today, more than in years past, it is recognized that "conscience needs knowledge". Science has liberated religion from falsehood, provided new means to do good, redefined many fundamental concepts in the heart of religion such us those of the neighbor and the relation of man to Nature, and revealed the beauty of the cosmos like no other means of knowing. Science, however, by constantly challenging the traditional cosmological claims of religion has weakened the inherent constancy of religious values and beliefs, and consequently their ability to moderate and stabilize societal changes largely induced by science. As Ian Barbour writes,[60] "When religion first met modern science in the seventeenth century, the encounter was a friendly one. Most of the founders of the scientific revolution were devout Christians who held that in their scientific work they were studying the handiwork of the Creator. By the eighteenth century many scientists believed in a God who had designed the universe, but they no longer believed in a personal God actively involved in the world and human life. By the nineteenth century some scientists were hostile to religion…. In the twentieth century, the interaction of religion and science has taken many forms".

Science has shown that it can be guided by values in the application of scientific knowledge, although much is still desired and much has happened that both science and society must be remorseful for. Values endure but they unavoidably change and often assume new meaning as new scientific knowledge is acquired.

A few other examples of the impact of science on values can be mentioned:

(i) *Science is called upon by society (the State) to help resolve moral issues*. In 1988, Zack[61] referred to a case where, as he put it, "the law wants science…. to tell it when human life begins, so that it may know when to define its ending as a crime…. The law wants to know if the zygote, embryo, and the fetus are human lives because it wants to know if these entities are entitled to the same rights and protections which the community has agreed to confer on human beings who have already been born. The issue is thus not whether the zygote, embryo, or fetus is human life in a scientific sense…. The issue is at what stage of development shall the entity destined to acquire the attributes of a human being be vested with the rights and protections accorded to that status. It is to the moral codes of the people that the law must turn for guidance in this matter, not to the arbitrary definitions of science." And he concludes: "To ask science to define human life in scientific terms for use by the law in moral terms is a travesty of both honorable traditions." The asking of the question is testimony not only to a profound misunderstanding of the capabilities and limitations of science, but of the difficulties involved.

(ii) *Science is called upon by society to address the moral issues posed by the growth of the power of science*. It has been repeatedly stated in this book that problems emerge when science begins to impinge on the autonomy of human beings and ethical issues are raised by profound distortions of humanness. It is widely acknowledged that possibly the principal ethical problem in the future is the control of man over his own biological evolution and the fact that such developments can escape the control of the scientist and society alike. Scientific knowledge especially in this area pushes man beyond the terms of all former ethics, although there may be cases where adjustment of values to scientific needs and vice-versa may be possible and would allow the benefit and avoid the ethical impasse.

(iii) *Issues of conscience*. Since WWII the frontiers of science and technology have become the frontiers of weaponry.[62] In this activity, ethical issues are of paramount importance and the role of the scientist is crucial. Many scientists have expressed deep concerns on several issues, although most continue to act "as usual" under different justifications. Interestingly, a striking similarity is seen[63] between nuclear science and genetic engineering: both major accomplishments confer on man a power for which he is morally unprepared. The physicist has learned this, the biologist, has not as yet. In the former instance once the scientists have provided the knowledge, others (e.g., the government and industry) took it and run away with it. Will this happen again in the latter case?

(iv) *Impact of science on values – emerging issues*. In the future, science will continue to confront society and to impact human values and ethics with increasing intensity foremost in such areas as the:

- Genome, genetic engineering, biotechnology;
- Internet, communications, robotics;
- Environment, climate change;
- Human dignity, poverty, rights to basic human needs such as food, water, and energy;
- Confrontation between frontier science and religion.

The growth of science and science-based technology will continue to impact traditional values. New norms and mechanisms embedded in the practice of science will be needed to moderate the adverse impact of science and science-based technology on society. The diverse issues eluded to in this Chapter have one thing in common: they all transcend science and they all transcend society. In their confrontation science meets conscience. Science and ethics become inseparable.

5.7 The Future of Values

Values vary with time and *loci* and this may be even true of the universal values we spoke of, because in practice they are subjectively perceived. The basic universal human values *are*, but they are, albeit slowly, constantly *becoming*. Recent scientific advances are seriously impacting human values, including the universal values we spoke of in this Chapter. The scientific challenge to values will never cease; there may be, as discussed earlier in this Chapter, new complementary values continuously emerging largely because of the advancement of science. There is universality and complementarity of values, but there is also a deep and continuous challenge to both by the emerging powers of science and science-based technology. The scientific challenge to the universal, common and complementary values and the pressure on them by the advancing science will continue unabated.

Values should not become speculative, but they should preserve their stability, standard and measure of value (αξία). For only then, they can secure a free society for man and science in which the integrity of science will be maintained and the dark side of science and science-based technology will be diminished.

References and Notes

1. Jérôme Bindé, in *Universal Values,* edited by Loucas G. Christophorou and George Contopoulos, Academy of Athens, Athens 2004, pp. 91–98.
2. Jérôme Bindé (Ed.), *The Future of Values – 21st-Century Talks*, Berghahn Books, New York 2004; UNESCO 2004 (ISBN UNESCO: 92-3-103946-6).
3. Momir Djurović (Ed.), Proceedings of the International Conference on *Values and 21st Century,* The Montenegrin Academy of Sciences and Arts, Podgoriga, Montenegro 2010.
4. Vernor Vinge, *VISION-21 Symposium*, NASA Lewis Research Center and Ohio Aerospace Institute, San Diego 1993.
5. The Millennium Project, *Future Global Ethical Issues*, 2009; State of the Future (12. Forecasts of Value Changes), pp. 106–109.
6. Huston Smith, in *Universal Values,* edited by Loucas G. Christophorou and George Contopoulos, Academy of Athens, Athens 2004, pp. 53–57.
7. Theodore T. Lafferty, *Nature and Values — Pragmatic Essays in Metaphysics*, University of South Carolina Press, Columbia, South Carolina 1976.
8. Loucas G. Christophorou, *Place of Science in a World of Values and Facts*, Kluwer Academic/Plenum Press, New York 2001, Chap. 1.
9. John Polkinghorne, in *Universal Values,* edited by Loucas G. Christophorou and George Contopoulos, Academy of Athens 2004, pp. 128–142.
10. Vladimir Lossky, *In the Image and Likeness of God,* St. Vladimir's Seminary Press, Crestwood, New York 1974.
11. Bentley Glass, in *Science, Technology, and Society — Emerging Relationships,* Rosemary Chalk (Ed.), The American Association for the Advancement of Science, Washington, DC 1988, pp. 37–43.
12. Edward O. Wilson, *On Human Nature,* Harvard University Press, Cambridge, Massachusetts 2004.
13. See, for instance, *Genes and Behavior*, Science **264**, 17 June 1994, pp. 1685–1739.
14. Gregory Stock, *Redesigning Humans – Choosing Our Genes, Changing Our Future,* Houghton Mifflin Company, New York 2003, pp. 2–4.
15. Neurological and psychological manipulation may control the development of human capacities, particularly those long considered most distinctly human such as speech, thought, choice, emotion, memory, imagination.
16. Leon R. Kass, *Life, Liberty and the Defense of Dignity — The Challenge for Bioethics*, Encounter Books, San Francisco 2002.
17. C. S. Lewis, *The Abolition of Man*, HarperCollins edition, San Francisco 2001.
18. Margaret Mead, *Culture and Commitment — A Study of the Generation Gap*, The American Museum of Natural History, Natural History Press, Garden City, New York 1970.
19. Loucas G. Christophorou, *Place of Science in a World of Values and Facts*, Kluwer Academic/Plenum Press, New York 2001, Chap. 9.

20. Scientism holds that the current content of science is the "final measurement of all truth and value".
21. Quoted by June Goodfield, in *Science, Technology, and Society — Emerging Relationships*, Rosemary Chalk (Ed.), The American Association for the Advancement of Science, Washington, DC 1988, p. 60.
22. The old fights between science and religion may be attributed in large part to religious views about the physical world that were not in agreement with the scientific facts. Such disagreements, whether in astronomy (over the heliocentric system), in geology (over the age of the Earth), in physics (over the uncertainty principle and free will), in biology (over the origin of the species and the special place of man in creation), led to misunderstandings and impacted negatively on both science and religion, especially the latter.
23. In an approach not unlike the anti-religious programs built into the educational system of the former Soviet Union during the 1950's, claiming "Science has disapproved religion!", a number of biologists recently use science to speak on religious matters. Stark examples are the books by Richard Dawkins, Edward Wilson, and Steve Pinker to mention just a few. In his book *The Selfish Gene* (see endnote 24), Dawkins claims that our existence is an artefact of chance and what counts is the survival of our genes and nothing else. Along the same line, Steven Pinker (see endnote 25) argues that consciousness is an illusion implanted in our brains because it gives better chance of passing our genes to the next generation. Similarly, evolutionary biologist Edward O. Wilson in his book *On Human Nature*s (see endnote 12) assures everyone that evolution is a myth that is now ready to take over Christianity. Such aggressive attacks by these and other scientists have been going on for some time and have prompted other scientists to ask: "Is evolution a secular religion?" (see endnotes 26, 27) Aggressive attacks on religion such as typified by Richard Dawkins in his book *The God Delusion* (see endnote 28) do not accurately represent the views of many other scientists (see endnote 29, 30), but unfortunately many in society accept the extreme hostility of a few scientists as representative of all scientists' view about faith. Indeed, Francis Collins (see endnote 31) protests that Dawkins "essentially discredits the spiritual beliefs of 40% of his colleagues as sentimental nonsense." Clearly, the proper boundaries of science must be observed.

24. Richard Dawkins, *The Selfish Gene,* Oxford University Press, Oxford 1989.
25. Steven Pinker, *How the Mind Works,* W. W. Norton & Company, New York 1997.
26. Michael Ruse, Science **299**, 7 March 2003, pp. 1523–1524.
27. Michael Ruse, *The Evolution — Creation Struggle,* Harvard University Press, Cambridge, Massachusetts 2005.
28. Richard Dawkins, *The God Delusion,* Houghton Mifflin, Boston 2006.
29. Alister McGrath and Joanna Collicutt McGrath, *The Dawkins Delusion?,* IVP Books, Illinois 2007.
30. Elaine Howard Ecklund, *Science vs. Religion — What Scientists Really Think*, Oxford University Press, New York 2010.

31. Francis S. Collins, *The Language of God*, Free Press, New York 2006.
32. Stephen Hawking and Leonard Mlodinow, *The Grand Design,* Bantam Books, London 2010.
33. The wisdom of religion bares no similarity to the "standard" of science. Although the language of religion differs from that of science an analogy may be drawn between the parable of religion and the model of science. Each parable describes a paradigm that complements that of the other parables by offering a particularly unique perspective on truth, on man and on man's behavior. As in science, acceptance of one paradigm does not mean giving up another. In science, for instance, the scientist accepted the paradigm of the quantum (photon) without giving up the paradigm of the wave; he needed both for a more complete description of physical reality. Similarly, in religion the Christian ethic is more complete with the paradigms of the parables of the Prodigal Son, the Good Samaritan, the Talents …, than by just the paradigm of only any one of them.
34. Stephen Jay Gould, *Rocks of Ages*, Ballantine Books, New York 1999.
35. Yet autonomous cultural values are increasingly on the defensive.
36. Loucas G. Christophorou and George Contopoulos (Eds.), *Universal Values,* Academy of Athens, Athens 2004.
37. Constantinos Despotopoulos, in Reference 36, pp. 27–45.
38. The Stoics were the first to introduce the value of human dignity and to defend the inherent worth of all people, laying the foundation of human rights.
39. We still have no scientific evidence of intelligent life anywhere in the universe. The existence of life elsewhere in the universe, however, is extraordinarily difficult to dismiss considering the tiny size of the planet Earth, the incomprehensible size of the universe and the recent discovery of a large number of planets in our Galaxy.
40. Loucas G. Christophorou, in Loucas G. Christophorou and Constantinos Drakatos (Eds.), *Science, Technology and Human Values,* Academy of Athens, Athens 2007, p. 473.
41. This is a new field dealing with ethical issues brought about by advances in biology, medicine, biotechnology and generally the manipulation of basic biology through altered *DNA*.
42. Felipe Fernandez-Armesto, in Reference 36, pp. 273–277.
43. Jüri Engelbrecht, in Loucas G. Christophorou and Constantinos Drakatos (Eds.), *Science, Technology and Human Values,* Academy of Athens, Athens 2007, pp. 217–223 and p. 477.
44. Human rights are natural rights in contradistinction to cultural rights. All human beings have human rights, which derive from the inherent human dignity; they are grounded in and are morally validated by universal values.
45. Loucas G. Christophorou, Reference 3, pp. 179–186.
46. Hans Jonas, *The Imperative of Responsibility*, The University of Chicago Press, Chicago, IL 1984.
47. Λουκάς Γ. Χριστοφόρου, *Επιστήμη και Αξίες*, Πρακτικά της Ακαδημίας Αθηνών, Τομ. 80 Α', Αθήνα 2005, σελ. 37–50, Αθήνα 2005. Loucas G.

Christophorou, *Science and Values*, Proceedings of the Academy of Athens Vol. 80 A', Athens 2005, pp. 37–50, Athens 2005.
48. Λουκάς Γ. Χριστοφόρου, *Βήματα στην Επιστήμη και τη Ζωή*, Σύλλογος προς Διάδοσιν Ωφελίμων Βιβλίων, Αθήναι 2009, σελ. 163–180.
49. Loucas G. Christophorou, *Place of Science in a World of Values and Facts*, Kluwer Academic/Plenum Press, New York 2001, Chap. 5.
50. InterAcademy Council (IAC), *Responsible Conduct in the Global Research Enterprise*, A Policy Report, September 2012.
51. All European Academies (ALLEA), *Ethics Education in Science*, September 2013.
52. All European Academies (ALLEA), The *European Code of Conduct for Research Integrity*, revised edition, Berlin 2017.
53. An Italian court accused and convicted seven Italian scientists of criminal negligence and manslaughter for failing to adequately warn residents before an earthquake struck the city of L'Aquila in central Italy in 2009 killing more than 300 people (see endnotes 54, 55). Subsequently, the Italian seismologists were cleared of these charges (see endnote 56). The incident illustrates both the misunderstanding of science by society and the excessive expectations of society from science, both a source of concern.
54. CBS / AP, October 22, 2012; CNN, October 23, 2012.
55. Stephen S. Hall, Nature **477**, 14 September 2011, pp. 264–269.
56. Alison Abbott and Nicola Nosengo, Nature/News, 10 November 2014.
57. See, for example, comments by Martin Rees and Jane Lubchenco, Science **327**, 26 March 2010, pp. 1591–1592; R. J. Cicerone, editorial, Science **327**, 5 February 2010, p. 624; Eli Kintisch, Science **327**, 26 February 2010, p.1070.
58. P. J. D. Drenth, J. E. Fenstad, and J. D. Schiereck (Eds.), *European Science and Scientists Between Freedom and Responsibility.* Luxembourg: Office for Official Publications of the European Communities, 1999.
59. Pieter J. D. Drenth, *The Universality of Scientific Values,* in Reference 36, pp.111–126.
60. Ian G. Barbour, *When Science Meets Religion*, HarperOne, New York 2000.
61. Brian G. Zack, in *Science, Technology, and Society — Emerging Relationships*, Rosemary Chalk (Ed.), The American Association for the Advancement of Science, Washington, DC 1988, pp. 98–99.
62. Loucas G. Christophorou, *Scientists and Society: Needs and Responsibilities*, Rend. Fis. Acc. Lincei **23,** 2012, (Suppl. 1), S23–S27; Λουκάς Γ. Χριστοφόρου, *Αέναη και Κρίσιμη Αλλαγή*, Πρακτικά της Ακαδημίας Αθηνών, Αθήνα 2009, Τομ. 84 Α', σελ. 109–132. *Perennial and Critical Change*, in Proceedings of the Academy of Athens, Athens 2009, Vol. **84**A, pp. 109–132.
63. L. F. Cavalieri, The Bulletin of the Atomic Scientists, December 1982, p. 82.

Chapter 6
Boundaries of Science

6.1 Introduction

The power of modern science to answer questions which can be defined scientifically is almost without limits (Chap. 3). Limitless also seems to be the spectrum of scientific questions to which scientific answers are sought; it stretches from the normally simple questions aimed at verification and systematization of scientific knowledge, to fundamental new questions of the scientific frontier where normally the answers are difficult and the scientific method is often at the limits of its capability. Scientific answers are also searched for fundamental questions that are generally assumed to have already been answered, as for instance, "*Is the speed of light in vacuum the maximum possible speed, the unsurpassable constant of nature*?" Questions and searches which largely demonstrate the way science works and evolves; the validity of scientific knowledge and the validation process by which it is established.

In this Chapter, I shall refer to two categories of scientific questions which lie at the boundaries of science. I have discussed such questions earlier[1] and lectured since on the subject on a number of occasions.[2, 3] The questions I wish to deal with in this Chapter can be formulated scientifically, but they have no scientific answer, they cannot be fully and adequately answered by science at the time they are posed, because at that specific time they lie beyond the capability of science; they are in principle "beyond the province of science[4]"; they belong to the area of *trans-science* (see endnotes 1–4). They demand special attention on the part of the scientist, who is often called upon to express scientific opinion without having adequate scientific data or knowledge on the subject to do so. Such questions fall in two categories:

Category 1 Questions in this category are defined scientifically but lie beyond the province of science when they are posed, without excluding the possibility of their scientific resolution in the *near future*. They can be split further into two subgroups, those that are purely technical (*Category 1A*) and those that have both technical and non-technical components (*Category 1B*).

L. G. Christophorou, *Emerging Dynamics: Science, Energy, Society and Values*,
https://doi.org/10.1007/978-3-319-90713-0_6

Table 6.1 Trans-scientific questions which when posed lie beyond the ability of science to answer, but their:

Resolution is not ruled out in the *near* future and *lie within science itself*	*Category 1A*
Resolution is not ruled out in the *near* future and *lie at the boundary of science on the one hand and the values and ethical norms of society on the other*	*Category 1B*
Resolution is beyond science for the *indefinite* future	*Category 2A*
Resolution is beyond science *possibly forever; teleological-type questions*	*Category 2B*

Category 2 Questions in this category are defined scientifically when they are posed, but they are beyond the province of science and are not expected to have a scientific answer for the *indefinite* future. As in Category 1, they can be split further into two sub-groups, those that are purely technical (*Category 2A*) and those that have both technical and non-technical components (*Category 2B*).

Trans-scientific questions require wisdom and breadth of scientific knowledge, for as Weinberg (see endnote 4) put it: "what the scientist can do in clarifying matters of trans-science differs from what he can do in clarifying matters of science." In the latter case, the scientist can bring to bear upon the question his scientific expertise to establish scientific truth, while in the former case he can at most "help delineate where science ends and trans-science begins". Similarly, there are interfaces of science with other human activities that can be characterized as trans-scientific, areas for instance between science and religion, science and philosophy, science and society, science and values. Earlier (see endnote 1), I have extended the concept of trans-science to include questions in those interfaces which can be stated scientifically, but – unlike those of Category 1 – transcend science possibly forever because they lie outside the scientific domain when they are posed and perhaps for the indefinite future. Table 6.1 summarizes the various categories and subcategories of trans-scientific questions to be discussed in the following sections.

6.2 Scientific Questions Without Scientific Answers

A. ***Examples of scientific questions which when formulated lie beyond the ability of science to answer, but their scientific resolution is not ruled out at some time in the (near) future and lie within science itself (Category 1A)***

First Example *The scientific verification and quantification of the possible health effects of low doses of ionizing radiation.* We are all exposed continuously to low doses of ionizing radiation; hence the scientific question: what is the lowest dose-level of ionizing radiation which causes damage to our health? Health physicists have excellent instruments for detecting ionizing radiation from radionuclides so that the resulting increment exposure can be determined even when it is a small fraction of the natural background radiation levels. However, the health impact of such very low levels of radiation exposure remains uncertain. At low dose levels, the possible health effects are negligible and their detection is difficult, and thus any possible correlation between cause and effect requires very large numbers of measurements and observations over a long period of time. An experimental or a theoretical answer to the question might become possible in the future through, for instance, the use of more sensitive measuring instruments or theoretical models of molecular radiation damage.[5]

Second Example *Possible health effects from low-frequency low-intensity electric and magnetic fields from electrical power transmission and distribution lines.* The problem is defined scientifically: electrical power transmission lines crisscross the space around us and consequently we are exposed to the low-intensity, low-frequency electric and magnetic fields which they produce. Hence the scientific question: what are the consequences of these fields for our health? Presently, there is no definite scientific answer to the question although there have been many credible studies[6, 7] which indicate no adverse health effects under the existing regulations and legal directives.[8]

Third Example What are the *possible health effects of the microwave radiation* which is emitted from various apparatuses to which we are exposed, or use daily, such as mobile phones? Although most studies conclude that there are no adverse health effects,[9] more data and relevant scientific knowledge may still be indicated for an unequivocal scientific answer. This question may still be in the area of trans-science.

Fourth Example *Genetically-modified foods.* Today, as was discussed in Chap. 2, it is possible to produce genetically-modified foods. There exists a huge and expanding industry of genetically-modified foods the quantity and variety of which increases on world markets, especially in certain countries such as the USA. Hence the scientific question: Are genetically-modified foods dangerous to human health and biodiversity, and does any such possible risk constitute sufficient reason to preclude their use even though they offer humanity the possibility to free millions of human beings from hunger? The industry of genetically-modified foods maintains that genetically-modified foods are safe. Others argue that further tests are necessary to ensure no possible long-range effects on human health and biodiversity. Genetically-modified foods are extremely important for society. It is therefore imperative that any questions relating to possible negative consequences of their use be addressed; some such questions may lie in the area of trans-science, awaiting more complete scientific answers in the future.

Other similar examples can be cited, and, undoubtedly, there will be new ones in the future.

An example, typical of questions in *Category 1B,* which lies at the boundary of science and technology on the one hand, and the values and ethical norms of society on the other hand, is: "*When does human life begin?*" At which phase ofits growth does the human embryo become a person with human rights and thus legal obligations toward it? In searching for scientific answers to this and other similar questions we are again confronted with the role of scientific knowledge in the values and the ethical norms of society and the impact of the latter in the execution of scientific research. Clearly, the need is again apparent for a serious and continuous dialogue between science and science-based technology and society in search for answers consistent with the scientific facts and needs on the one hand, and human values and ethics on the other.

The examples which were mentioned thus far lie at the boundary of science and technology on the one side, and the political and social interests and values on the other. In their scientific and social accommodation and resolution, often enters the evaluation of risk vs. benefit of the scientific technology, the assessment of which is normally done based on the available scientific data and the need of society to decide the issue at hand in order, for instance, to enact new laws and regulations. Although often not all required scientific facts and data are available, the need for an immediate decision on the part of the State burdens the scientist to offer an opinion before the question can have its final scientific resolution. It befalls on the scientist to bring to the discussion whatever scientific knowledge he / she has on the subject and to avoid drawing conclusions beyond those allowed by the existing scientific facts. In such cases, it is the duty of the scientist to delineate the boundary where science stops and trans-science begins.[10]

Naturally, at any given time, there are many purely scientific questions which await scientific answers. These are not trans-scientific questions, but rather questions in the normal course of an evolving science. A couple of examples will suffice to make the distinction clear. (i) About a century ago, conventional wisdom held that the universe was static and eternal. In fact, Einstein not only ridiculed Georges Lemâitre for suggesting a beginning, but also invented the cosmological constant for the purpose of allowing a static universe.[11] Today most scientists accept a cosmic beginning (Chap. 1). (ii) About 85 years ago, observes Gingerich,[12] we did not know how to answer scientifically the question "*Why is iron so much more common than gold and uranium*?" But, as he writes, "within three decades it became a perfectly sensible question"; astronomers, he writes, "began to outline how elements form in stars, how iron represents an end point in the evolution of normal stars, and how the precious heavy elements in particular result from rare supernova explosions that generate these atoms in a swift fiery shower of neutrons and then spew them out into space ... gold and uranium are the stuff of supernovas." (iii) Today, the scientific question "*What is the origin and nature of dark energy?*" is still open. The question "*Are we alone in the universe or is there extraterrestrial life?*" is still unanswered. Science can neither state for sure that we are the only self-conscious, thinking creatures in the universe, nor, if it turns out that life is restricted to Earth, that it is an historic accident. These are questions science cannot fully answer now but hopes to do so at an unspecified future time.

B. ***Examples of scientific questions which when formulated lie beyond the province of science and their scientific resolution is beyond science possibly for ever (Category 2)***

Trans-science questions in *Category 2* that can be stated scientifically but cannot be answered by science because they lie beyond its province are those at the boundary of science and philosophy, science and metaphysics, and science and religion.

First Example Ultimate, origin questions and deep questions of existence, such as "*What is the origin of life?*" fall in this category. The origin of life cannot be separated from our own existence; we are here and hence some sequence of events must have led to us. The question "What is the origin of life?" is naturally followed by a series of other scientific questions: "How did life evolve from inorganic matter?"; "How did the first living organism originate?"; "How from dead atoms and molecules emerged conscious human beings?"; "What is the scope of life?" and so on. These are fundamental existential and teleological questions, which are scientifically unanswered. Part of the difficulty in answering such questions scientifically is the unimaginable complexity of living organisms.

The origin of life remains a fundamental scientific question which reductionist science cannot answer. Scientists from various branches of science offer opinions as to how life began, but they have no scientific answer. At the 30th General Assembly of the International Council for Science (ICSU) held in Rome in September 2011, Professor Werner Arber, Nobel prize winner in Physiology or Medicine, speaking on the topic "Updated scientific knowledge on biological evolution" repeated what others[13, 14, 15] declared earlier, namely, that they do not understand the origin of life and that the existence of life must be considered an elementary fact that cannot be explained!

Obviously, there are scientists who claim that it is just a matter of time until science discovers the sequence of chemical reactions which led to the emergence of the first single-cell organism on Earth. In fact, the materialist's view of life holds that life has essentially been reduced to the level of physics and chemistry and that everything, including man, can be reduced to their constituent atoms and molecules. And, conversely, the materialist's view of life holds that reductionist science can lead from molecules to some initial single-cell organism which emerged from inorganic matter and in time led to us.

But, how did the first self-organizing and self-replicating organism which is the carrier of coded information and constitutes the evolutionary link between dead matter and living organisms emerge from inorganic matter? Even if we accept the emergence of the initial single-cell organism from the atoms and molecules of matter, still the question as to how life arrived from the single-cell organism to man remains unanswered. To scientists studying the origin of life, the question is not whether life could have originated via chemical processes involving non-biological components, but rather which of the many pathways might have been followed to produce the first cells. It is thus often being asked: "Will science ever be able to identify the path of chemical evolution that succeeded in initiating life on Earth?" Perhaps, but even when a living cell were to be made in the laboratory it would not prove that Nature followed the same pathway billions of years ago. Clearly, a long path leads from the origins of primitive life 3.5 billion years ago to the diversity of life today[16, 17]. Reductionist science, at least for now, has no answer as to how unconscious atoms combined to give rise to intelligent conscious human beings. We presently know neither how primitive life emerged nor do we know how it led to us, admit many scientists. Life can neither be reduced to the properties of the constituent parts of an organism, nor can the conscious self be explained by the Darwinian evolutionary process[18]. Deep questions of existence belong, at least for now, to trans-science.

Second Example "What is the origin of the universe?", "What caused its beginning?", "Why did the universe begin in an explosion?", "What came before the beginning?" are questions which remain unanswered. As was discussed in Chap. 1, today science has strong indications that the universe did not always exist but it had a beginning 13.8 billion years ago in a cosmic explosion. "How and why did the cosmic explosion take place?", scientifically we do not know. "What existed before the cosmic explosion?", again scientifically we do not know. "What physical laws were valid then and from where did those 'laws of Nature' come?", again scientifically we do not know. Even the question "What is it that really exists?" is scientifically unanswered; Hawking's answer[19] that "reality depends on your model" is clearly not adequate.

Third Example *Questions which lie outside the province of science when they are formulated and their scientific resolution is beyond science possibly forever.* Classic examples are the ontological- and teleological-type questions. Such questions are in the domain of philosophy and religion rather than in the domain of science. They thus are not possible to answer scientifically. "Why was the universe brought into being?", "Does the universe have a purpose?", "Is there ultimate meaning to human existence?", or "Why is the universe comprehensible?" It might be instructive to mention a few statements made by distinguished scientists regarding the last question: Einstein's often mentioned quote that "The most incomprehensible thing about the universe is that it is comprehensible"; Steve Weinberg's quote that "The more the universe seems comprehensible, the more it also seems pointless[20]"; and quotes by others that "The universe might just be comprehensible because it is part of its purpose to be so[21]". Wigner pointedly also asked[22] "Where in the Schrödinger equation do you put the joy of being alive?" Questions of this type can be considered "ultimate", "unanswerable", "unprovable"; they lie beyond the scope of science.

6.3 Extrapolation of Scientific Knowledge

Not all knowledge can be considered "real" and "objective", and not what is not known is necessarily nonexistent. When science arrives at its boundaries, as in the case of trans-scientific questions, there may be cases where we can be led beyond science by cautiously extrapolating from that which we scientifically know to what we do not know, fully recognizant that in so doing we are not doing science, but we are extrapolating from it. The scientist is obviously free to extrapolate his scientific knowledge and locate himself in trans-science assuming that he makes abundantly clear that even if his extrapolation is based on scientific knowledge, it remains an extrapolation beyond the validity and validation process provided by science, and it lies in a different domain of knowledge.

Let us give an example and ask: "What is energy and from where did the *first (primordial)* energy come?" If energy and matter evolve according to the laws of nature, from where did those laws come? In searching for scientific answers to those questions we realize that science is not able to tell us what energy is (although scientifically we know many things about energy[23, 24]) or explain how the initial energy came from nothing. Even the suggestion that the quantum vacuum can constitute a source of energy, presupposes the existence of the quantum vacuum itself and the laws that determine its behavior.

Let us then extrapolate and move beyond science into the philosophical dimension of energy (see Appendix) to the boundary between science and philosophy.

For many centuries, energy (matter) was considered infinite in space and time. Aristotle considered matter eternal (αἰώνια), imperishable (ἄφθαρτον), and unborn (ἀγέννητον); *«ἄφθαρτον καὶ ἀγέννητον ἀνάγκη αὐτὴν εἶναι»*, he wrote[25]. However, if the universe started abruptly at a moment, this moment is the beginning of time and there was nothing before it. Therefore, the abrupt transition from absolute nothing to the world means that the energy at the beginning of the world did not always exist, it did not precede the world; it means that energy is neither eternal nor infinite.

Aristotle referred particularly to "***prime matter***"[26, 27] (***πρώτη ὕλη***), which he considered as the substratum (ὑποκείμενον) of all things and all change, the substratum of "*all energetic beings*" (*«ὅλων τῶν ἐν ἐνεργείᾳ ὄντων»*). According to Aristotle, the prime matter contains all form (μορφή, εἶδος) potentially (ἐν δυνάμει). The transition from the prime matter to form, from the potential to the energetic being (*«ἀπό τὸ ἐν δυνάμει εἰς τὸ ἐν ἐνεργεία* ὄν»), is the perpetual motion (ἀέναη κίνησης) that takes place between them (see endnote 27)[28, 29]. If, then, we consider that the prime matter of Aristotle, corresponds to the initial, primordial energy at the beginning of the universe, and if we further consider that Aristotle's motion originates from the perennial change caused by the forces of nature, we recognize that parts of the Aristotelian philosophy are relevant to the modern scientific view. The initial energy and all its subsequent forms, through the physical fields and the forces they generate, are transforming incessantly the universe, and the transformations of energy cause the perpetual change of the physical world and its evolution, and consequently, the transition from the "potential" to the "energetic" being (*«ἀπό τὸ ἐν δυνάμει εἰς τὸ ἐν ἐνεργεία* ὄν»). All material reality is possible – potentially – as argued by Aristotle, and comes into existence, into energetic beings (*«ἐν ἐνεργεία* ὄντα»), over time.

Let us then accept a beginning of energy, time, space and change, the explanation of which is beyond science and, in contrast to Aristotle, but according to modern science, let us accept that the energy-matter is neither infinite nor eternal. Then we can consider that the initial energy is the source of the initial fields and forces that shaped the early universe. Thereafter, all respective forms of energy – through the physical fields and the forces they produce – are transforming perpetually the energy and lead to the perennial change and evolution of the universe, to its current form. Both, modern science and "old" philosophy converge on the idea of *a transient physical world incessantly changing and evolving via the unceasing transformations of energy.*

It can be concluded from the trans-scientific questions mentioned in the preceding sections of this Chapter that:

- There are limits to the potentialities of science, limits which although continuously changing, or are removed by the evolving science, some such limits remain and new ones emerge.
- It is necessary to recognize the significance of the scientifically unanswered questions and to delineate the boundary where science stops and trans-science begins, especially when the scientist is called upon to offer whatever relevant knowledge he / she is in possession of concerning the question.
- As scientists and as citizens, we have the obligation to postpone providing answers to even "hot" trans-scientific questions when there is no adequate scientific evidence to justify scientific answers.
- We can of course extrapolate scientific knowledge beyond the strict scientific domain assuming that we are clearly distinguishing when we are within science and when we are extrapolating beyond science.

6.4 Examples of Extreme Cases of Scientific Reductionism and Inductive Extrapolation Therefrom

The reductionist method of science proved a powerful method of knowing (Chap. 3). By reducing nature to its most extreme fundamental level and by learning "all" that can be learned at that elemental level – the properties and behavior of microscopic matter – the scientist built inductively step-by-step his knowledge of the macroscopic physical world including life itself. However, as we have discussed in Chap. 3, this reduction-induction process, is not without limits. We shall illustrate the limits of this process by referring to two cases of extreme reduction-induction: one in physics, the other in biology.

Example from Physics *From the fundamental constants of physical science to man?* Let us first look at the so-called fundamental constants of Nature. "Gradually", wrote physicist John Barrow[30] "we have identified a collection of mysterious numbers which lie at the root of the consistency of experience. Despite the incessant change and dynamic of the physical world, there are aspects of the fabric of the Universe which are mysterious in their unshakable *constancy*... There is a golden thread that weaves continuity through Nature. It leads us to expect that certain things elsewhere in space will be the same as they are here on Earth; that they were and will be the same at other times as they are today… they lie at the root of sameliness in the Universe." These are the constants of Nature (e.g., the speed of light in vacuum c, the gravitational constant G, Planck's constant h, the electric constant ε_0, the elementary electrical charge e[31,32,33]), which "capture our greatest knowledge and our greatest ignorance about the universe, … for while we measure them to ever greater precision, fashion our fundamental standards of mass and time around their invariance, we cannot explain their values" (see endnote 30). Indeed, the constants of Nature are peculiar wrote Wilczek[34] in that several of them are exceedingly large or exceedingly small, while others[35] wonder whether the 'constants' have indeed stayed constant through the entire period of time over which the universe itself has changed.

A number of scientists[36] view the constants of Nature and the theoretical descriptions they underpin, entirely as artefacts of a particular human choicc of representations to make sense of what is observed. Alternatively, many scientists argue (see endnote 30)[37] that if their values were different from what they are, then the necessary conditions for the emergence of life based on the carbon atom would not be satisfied and we would not be here. Conversely, our existence, it is said, explains why those constants of Nature have the values they do. This in essence represents the so-called Anthropic Principle: *from the fundamental constants of Nature to us.* There are various versions of the Anthropic Principle (see endnote 37)[38,39], which essentially introduce a teleological explanation of the constants of Nature: the universe has been tune with the sole *purpose to allow the emergence of conscious beings like us, and this is the reason the constants of Nature have the values they do.*

The process of extrapolation from the constants of Nature to man, represents an extreme form of reduction-induction.

Example from Biology *From random genetic changes to the evolution of all life?* Reductionism, so familiar in physics, is gradually being extended to biology. Just as physical science has shown that at the atomic level there is similarity in composition and function of the physical world, modern molecular biology and genetics have shown that at the macromolecular level there is similarity in composition and function of all organisms; the basic macromolecular structures of organisms are fundamentally the same and remained so for hundreds of millions of years.

The study of many organisms from bacteria to humans has shown that the genetic code through which information coded in the *DNA* and *RNA* molecules is transmitted to the progeny of organisms is common for all life. Comparisons of the genes of various organisms has shown their degree of similarity and evolution over time. The occurrence of mutations in the *DNA* molecule causes changes in the genes of organisms which can be inherited; consequently, the *DNA* molecule contains information about the history of evolution of the various organisms. Natural-radiation-induced random changes at the molecular level lead to random errors in the self-replicating of the *DNA* of organisms and those errors, it is argued, affect the evolution of organisms.

Many scientists however, argue that such random errors in the auto-copying of the *DNA* structure cannot explain the enormous variety of living organisms if one considers the expected low number of mutations resulting from the action of natural radiation and the fact that the preponderance of those mutations is damaging rather than beneficial[40,41].

If the evolutionary mechanism of living organisms is based on genetic changes at the molecular level, it is based on physicochemical phenomena and processes at the microscopic level. *The process from the random microscopic phenomena to the extreme macroscopic phenomenon of evolution of all life constitutes an extreme case of reduction-induction.*

Similarly, the reduction-induction process *"From atoms to molecules, from molecules to a single-cell organism, and from a single-cell organism to the human"* assumes that all living organisms have come from an initial single-cell-organism which emerged from inorganic matter. How did this happen? We simply do not now know. What we do know is that even the simplest organism is unimaginatively complex. Even the level of complexity of the *DNA* macromolecule is so horrendous, that one wonders how it could have been possible to emerge from inorganic matter automatically. According to Mitchell Waldrop,[42] biologist Stuart Kauffman was so amazed at the vastness of complexity of biological macromolecules that wondered how molecules such as proteins could be formed from smaller molecular units by chance. Many people, he writes (see endnote 42), attempted to calculate the probability for something like this to happen, and their answers were similar: If the making of these macromolecules is a result of chance, one ought to have waited for times very much longer than the age of the universe for the creation of even one useful protein molecule, let aside producing all the other compounds for a functional cell. If the origin of life is the result of chance, he concluded, "it is truly a miracle". *Thus, one is obligated to conclude that the notion that inductively we can be led from inorganic matter to organic molecules and to the single-cell organism and from it to Homo sapiens represents an extreme form of reduction-induction.*

References and Notes

1. Loucas G. Christophorou, *Place of Science in a World of Values and Facts*, Kluwer Academic / Plenum Publishers, New York 2001.
2. Λουκάς Γ. Χριστοφόρου, *Βήματα στην Επιστήμη και τη Ζωή*, Σύλλογος προς Διάδοσιν Ωφελίμων Βιβλίων, Αθήναι 2009, σελ. 197–202.
3. Λουκάς Γ. Χριστοφόρου, *Στα Όρια της Επιστήμης: Επιστημονικά Ερωτήματα Χωρίς Επιστημονικές Απαντήσεις,* Ίδρυμα Ιατροβιολογικών Ερευνών της Ακαδημίας Αθηνών (ΙΙΒΕΑΑ), 12 Δεκεμβρίου 2011. Loucas G. Christophorou, *At the Boundaries of Science: Scientific Questions Without Scientific Answers,* Foundation for Biomedical Research, Academy of Athens, 12 December 2011.
4. Alvin M. Weinberg, Minerva **10**, April 1972, p. 209; Science **177**, 21 July 197, p. 211.

5. Jocelyn Kaiser, Science **331**, 25 March 2011, p.1504; see discussion on the subject in Physics Today, May 2000, pp. 11–15 and 76 (radiation risks and no-threshold theory); also, in Physics Today, July 2016, pp. 10–16 (the linear no-threshold theory: readers weigh in).
6. NIEHS Report, *Assessment of Health Effects from Exposure to Power-Line Frequency Electric and Magnetic Fields*, National Institute of Environmental Health Sciences of the National Institutes of Health, NIH Publication No. 98-3981, August 1998.
7. See, also, subsequent articles and reports, for instance, in British Journal of Cancer and American Journal of Epidemiology.
8. Δημήτριος Κ. Τσανάκας, *Ενέργεια και Περιβάλλον*, Επιτροπή Ενέργειας της Ακαδημίας Αθηνών, Αθήνα 2008, σελ.181–192. Βλέπε, επίσης, Ευάγγελος Λεκατσάς, σελ.193–201. D. Tsanakas, in *Energy and Environment,* Energy Committee of the Academy of Athens, Athens 2008, pp. 181–192; see, also, E. Lekatsas pp. 193–201.
9. See relevant articles in Health Physics, Journal of Exposure Analysis and Environmental Epidemiology, Occupational and Environmental Medicine, American Journal of Epidemiology and British Medical Journal; see, also, http://en.wikipedia.org/wiki/Mobile_phone_radiation_and_health
10. Unfortunately, there have been instances in the past where this was not followed; for example, in the case of the tobacco industry (effects of smoking on human health).
11. Lawrence M. Krauss, *A Universe from Nothing*, ATRIA paperback, New York 2012.
12. Owen Gingerich, *God's Universe*, The Belknap Press of Harvard University Press, Cambridge, Massachusetts 2006.
13. Max Perutz, *Is Science Necessary?*, Oxford University Press, Oxford 1991.
14. Paul Davies, *God & the New Physics*, Simon &Schuster, New York 1983.
15. Henry Margenau and Roy A. Varghese (Eds.), *Cosmos, Bios, Theos*, Open Court, La Salle, Illinois 1992.
16. There are of course related questions which science is close to providing answers. For instance, the study of genomes of various species leads to the conclusion that humans share a common ancestor with other living organisms (see endnote 17).
17. Francis S. Collins, *The Language of God*, Free Press, New York 2006.
18. John C. Eccles, in Timothy C. L. Robinson (Ed.), *The Future of Science – 1975 Nobel Conference*, John Wiley & Sons, New York 1977, pp. 98–101.
19. Stephen Hawking and Leonard Mlodinow, *The Grand Design*, Bantam Press, London 2010, p. 16.
20. Steven Weinberg, *The First Three Minutes*, Basic Books, New York 1993, p.154.
21. Reference 12, p. 96.
22. Eugene P. Wigner in an interview by István Kardos, *Scientists face to face*, Corvina Kiadó 1978, p. 370 (ISBN: 963130373X).

23. Loucas G. Christophorou, *Energy and Civilization*, Academy of Athens, Athens 2011.
24. See, also, Chap. 1 and Appendix.
25. Aristotle, *Physics*, 192a28-29.
26. Aristotle, *Physics,* 192a32-34.
27. Δήμητρα Σφενδόνη-Μέντζου, «Η Αριστοτελική πρώτη ύλη μέσα από το πρίσμα της Κβαντικής Φυσικής και Φυσικής Στοιχειωδών Σωματίων», in *Ο Αριστοτέλης σήμερα. Πτυχές της Αριστοτελικής Φυσικής Φιλοσοφίας υπό το πρίσμα της σύγχρονης επιστήμης,* Σελ. 61–107, Εκδόσεις Ζήτη, Θεσσαλονίκη 2010. Demetra Sfendoni-Mentzou, *Aristotle Today: Aspects of Aristotelian Natural Philosophy in the Light of Modern Science*, Ziti, Thessaloniki 2010.
28. Aristotle, *Metaphysics*,1069b 35.
29. Aristotle, *Metaphysics,* 1072a19-1072b30. http://users.uoa.gr/~nektar/history/tributes/ancient_authors/Aristoteles/metaphysica.htm
30. John D. Barrow, *The Constants of Nature*, Vintage Books, New York 2004.
31. https://en.wikipedia.org/wiki/Physical_constant
32. E. Richard Cohen and Barry N. Taylor, *The Fundamental Physical Constants*, Physics Today, August 1999, pp. BG5– BG9.
33. Keith A. Olive and Yong-Zhong Qian, Physics Today, October 2004, pp. 40–45.
34. Frank Wilczek, Physics Today, January 2006, p.10.
35. Martin Rees, *Before the Beginning – Our Universe and Others*, Basic Books, New York 1998, p. 224; Keith A. Olive and Yong-Zhong Qian, Physics Today, October 2004, pp. 40–45.
36. David Bohm, *Wholeness and the Implicate Order*, Routledge & Kegan Paul, New York 2005.
37. John D. Barrow and Frank J. Tipler, *The Anthropic Cosmological Principle*, Oxford University Press, Oxford 1986.
38. If the values of the fundamental constants of Nature differed significantly from those they actually have, then the necessary conditions for the emergence of carbon-based life would not be met, and we would not now exist.
39. A weak and a strong version of the anthropic principle is normally distinguished. According to the *Weak Anthropic Principle* our existence is the explanation of why the constants of Nature have the values they do. *The Strong Anthropic Principle* states that the universe is fine-tuned for the very purpose of allowing human beings with consciousness and subjective awareness to evolve (The universe must have those properties which allow life to develop within it at some stage in its history.).
40. Michael J. Behe, *The Edge of Evolution – The Search for the Limits of Darwinism*, Free Press, New York 2007.
41. The role of random mutations is in fact considered to be the weakest point of the hypothesis that random mutations and subsequent natural selection led to the differences between species and their differentiation.
42. M. Mitchell Waldrop, *Complexity – The Emerging Science at the Edge of Order and Chaos*, Simon & Schuster Paperbacks, New York 1992; Stuart A. Kauffman, *Reinverting the Sacred*, Basic Books, New York 2008.

Chapter 7
Energy

7.1 Introduction

Surprisingly, while the word "energy" (ενέργεια) was introduced by Aristotle in the fourth century BC and entered the philosophical/theological debates of Christianity in the fourth century AD (see Appendix), in scientific/technological sense, the word "energy" did not exist[1] until 1807 despite its fundamental role in science. Thomas Young (1773–1829) was the first to introduce the word energy into physics and to recognize its scientific significance.[2] The scientific principles that govern energy were not established[3] until the mid1800s and even those principles had to be modified when it was discovered that mass was a form of energy. Energy has been usually defined as the equivalent of, or as the capacity for, doing work.

We have referred to energy and its fundamental significance for society and life in general in earlier chapters of this book. In this chapter we focus on past, present, and future energy sources, and on the energy needs of modern society and science-based technology.

7.1.1 The Fundamental Role of Energy

I have outlined the fundamental role of energy in science and stressed the essential role of energy for every aspect of life in a book on Energy and Civilization.[4] All ordering and organization in the universe depends on energy, its transformations on the microscopic and on the macroscopic level, and its flow. The flow of energy through a system acts to organize that system[5] and the total amount of work a system can do depends on the energy flow through the system and the diversity of the work done. An isolated system cannot do steady-state work. A continuous net flow of energy through a system (from and to an external reservoir) is necessary for the system to be maintained in a steady state and to avoid drifting toward equilibrium.

L. G. Christophorou, *Emerging Dynamics: Science, Energy, Society and Values*,
https://doi.org/10.1007/978-3-319-90713-0_7

Maintaining order requires continuous work, which can be supplied only by the flow of energy from a source to a sink. Millennia have passed before man has discovered and appreciated the fundamental relation of energy and life.

7.1.2 Energy for Civilization

When Thales of Miletus (sixth century BC) was asked "which thing is the mightiest, the strongest, of all?" («τί ισχυρότατον;»), he replied *"need"* (*«ανάγκη»*[6]). We and every other form of life on this planet are inextricably linked to and depend on energy. Today's society, more than any other society in history, needs energy. The standard of living of every nation is rooted in energy and no other nation exemplifies this fact better than the United States of America. Today, as in the past, what the world needs most is *energy*. Energy, today, defines and constrains progress; it is a requisite of economic prosperity and civilization, including democracy and freedom. The greatest choice to the average citizen in human history, is today available in the high-energy-consuming societies of the developed countries of the world.

Man's need for energy sources that were independent of time and place led to the discovery of new energy sources and none was discarded; every form of energy that had been discovered it was used; each new source of energy supplanted rather than replaced those that came before.

History is replete with examples of how man has used energy to create civilization and of how civilization took a downward trend when society failed to control the consequences of energy demands that exceeded the limits of the available energy sources.[7,8,9,10,11,12] A reduction of per capita energy consumption has, in the past, led to a decline in civilization and a reversion to a more primitive way of life, as, for instance, in India, China, and the Middle East (see endnotes 8, 9).[13]

Historically, the Sun has been man's source of energy; it provides heat and light for photosynthesis and affects the motion of the wind and the water. Ancient high cultures such as in Greece and Rome used solar energy, wind energy, muscle energy from humans and animals, and wood fuel for their energy needs. Abundant energy from those sources was essential for their democracy and social development. The recent discovery of energy sources such as coal, oil, gas and uranium, enabled unprecedented human mobility, industrial mass production, electrification, appliances, communication systems, and so on, which allowed better living conditions for an increasingly larger number of people, and caused near-exponential growth of population.[14] The economic growth over the last 200-years or so, named "the energy revolution", became possible by access to cheap and abundant fossil fuels (mostly coal) to which country-after-country turned to for its development; it accustomed humanity to these seemingly abundant energy sources available at reasonable cost and in concentrated and easy-to-use forms, but finite, hazardous and highly polluting energy sources. Fossil fuels are often taken for granted; their high-consumption rate is non-sustainable and clearly demands new "clean" long-term alternatives.[15] What sources of energy can be used for sustainable development? ALL, prudently!

The annual power consumed collectively by humanity exceeds the equivalent of 14 terawatt (TW) and is mostly provided by fossil fuels.[16,17] These amounts of energy must be replaced by carbon-free energy sources, but most energy experts predict that due to global population growth and economic development, consumption levels may reach 30 TW by 2050 provided mostly by fossil fuels.

7.2 Primary Energy Sources

7.2.1 Energy for Today and for Tomorrow

The argument is often being made that modern civilization needs to stabilize population growth and energy consumption and to abandon the notion of continued growth, for neither economic progress nor societal complexity and size can go on increasing indefinitely. *Sustainable civilization requires sustainable development, and sustainable development needs sustainable energy sources.* Sustainable energy sources are anticipated to extend far into the future and to support life in the long run. Sustainable energy, also, means wise and prudent use of energy, clean and efficient technology, and energy sources compatible with the environment.

Until recently, the world energy consumption was minor compared with the amounts of the available energy sources. This is so no more. Availability of clean, safe, secure, and affordable energy is one of the greatest challenges facing the world today. Future human prosperity would likely depend on how successful society will be in securing the supply of reliable and affordable energy it needs, and in effecting a rapid transition to a low-carbon energy supply. Future strategies to cope with energy demands on the one hand and with the environmental/climate change consequences of energy production and use on the other hand, will include almost every available technology. Sustainable solutions to the problems of the energy-environment-climate change require an understanding of the basic problems involved, as well as the phenomena which are relevant to both energy and the environment/climate change. It would also require development of new energy technology and use of novel materials. The idea, for instance, is being articulated for an energy system equivalent of the Internet, a sophisticated web of technologies/information systems/services capable of drawing electricity from where it is abundant and sending it to where it is needed, thus replacing the need for extra new plants, and responding to personal energy needs and demands. Increasingly, sharing natural resources will require close international cooperation, security, and peace.

The main *primary* sources of energy at man's disposal today fall into four categories:

- *Fossil fuels*, mainly coal, oil and natural gas;
- *Renewable energy sources*, mostly hydroelectric, solar, wind, geothermal, biomass and tidal;
- *Nuclear power*, from nuclear fission and in the future from nuclear fusion; and
- *Energy conservation.*

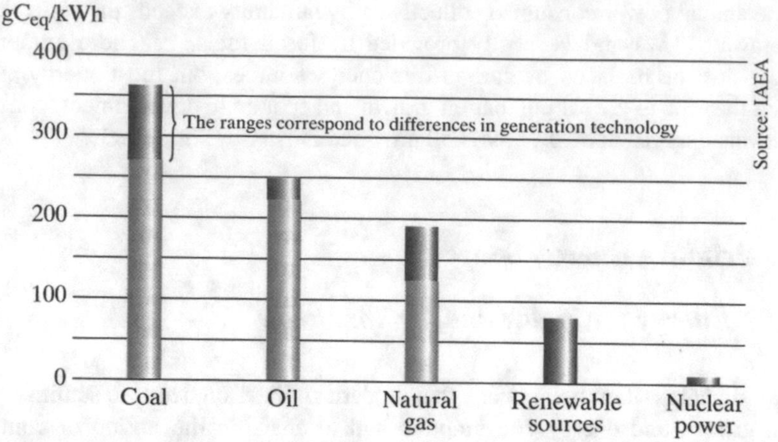

Fig. 7.1 Total emission of greenhouse gases in the production of electricity from fossil fuels, renewables, and nuclear power (see endnote 18)

Each of the first three primary energy sources has its advantages and its disadvantages. Each has its price and each has its degree of danger which is largely connected with the associated technology. Let us then refer briefly to these main categories of primary sources of energy and inquire as to their potential contribution to modern society's energy needs and their corresponding environmental/climate change impact.

The overall greenhouse gas emissions in the production of electricity from fossil fuels, renewables and nuclear power are compared[18] in Fig. 7.1. Figure 7.2 shows that within the fossil fuel category used for electricity generation and for transport there is considerable variation in the "carbon footprint",[19,20] and thus serious consideration should be given to the kind of fossil fuel used for both electricity generation and, especially, for transport. In Fig. 7.1, the total amount of greenhouse gases emitted to the environment for electricity generation from the various energy sources indicated, is given in units of gram-carbon-equivalent per kWh (gC_{eq}/kWh).

7.2.2 *Fossil Fuels; Mainly Coal, Oil and Natural Gas*

The most crucial element regarding fossil fuels is the fact that modern industrial and technological civilization is still dependent on the energy provided by those cheap, concentrated, storable and transportable fuels; about 80% of the current world-wide energy needs comes from fossil fuels. The burning of those fossil fuels, especially coal, is the most damaging to the environment.[21]

Fossil fuels are not renewable energy sources; they can be used only once. The main fossil fuels are coal, oil and natural gas.

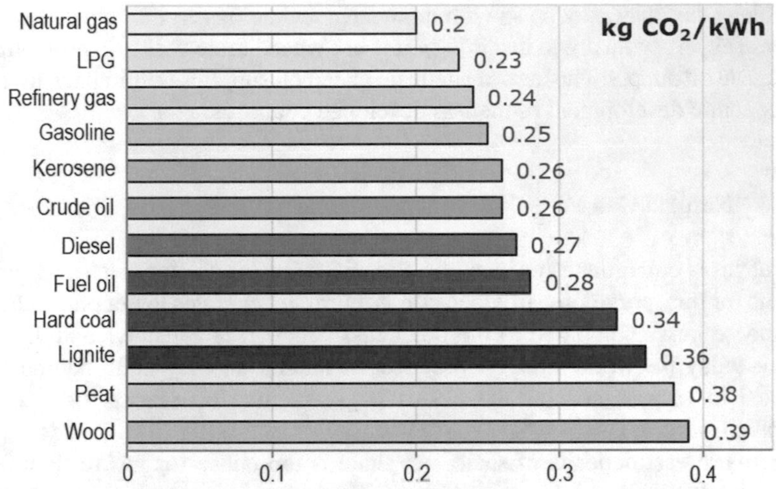

Fig. 7.2 Emissions of CO_2 (kg CO_2/kWh) generated from various carbon sources. (Reproduced with the permission of Volker Quaschning) (see endnote 19)

7.2.2.1 Coal

The use of coal is expected to continue and possibly to increase due mainly to the rapid world-wide demand for electricity and the large reserves of coal compared to other fuels. Even if the burning of fossil fuels is limited to the now known reserves, experts caution that the consequences of burning those fuels (without capture and sequestration of CO_2) will double the amount of CO_2 in the atmosphere. *Consequently, the substantial use of coal in the future is incompatible with efforts to stabilize CO_2 concentrations in the atmosphere and carries with it unacceptable risks for climate change.* Clearly, modern technologies are needed which will allow the use of fossil fuels in ways which are compatible with the reduction of environmental and climate-change risks. Indeed, it is primarily the responsibility of countries whose development and supremacy depend largely on coal (e.g., China, India and the USA) to conduct research in better combustion and more effective capture and storage of CO_2.[22,23,24]

7.2.2.2 Oil

Oil is broadly defined as conventional and unconventional. Most (~90%) of all oil to date is conventional type; unconventional oil refers to ultra-deep-water oil, tar sands and shale oil and is more expensive.[25] Except for Russia, the Asian countries China, India and Japan are not endowed with energy resources, especially oil. According to Hubbert,[26] oil is the most important form of energy in the world today; no other energy source equals oil's intrinsic qualities of extractability, transportability, versatility and cost. There are types of oil which are not recoverable with present technology, but they may be so with innovative technology in the future. However, there are experts who argue that it is "not if but when" there will be a problem with inadequate oil supply. The lack of adequate cheap oil supplies will impact adversely the economic development of the less developed countries.

7.2.2.3 Natural Gas

Natural gas is emerging as an energy source with greater use for electricity generation and for transportation partly because its burning generates lower percentages of greenhouse gases compared to the other fossil fuels (see Figs. 7.1 and 7.2). The USA is today the world's largest producer of natural gas. By 2035 natural gas is expected to surpass coal as the USA's largest source of energy for electricity generation.

There are vast deposits of shale, but shale is too dense for gas to flow freely. Horizontal drilling and hydraulic fracturing (hydrofracturing or fracking) is used to extract oil and gas from shales.[27,28,29,30,31] Estimates by IEA[32] put the USA shale gas deposits at *16 trillion cubic meters*. This source of energy has been intensely explored in the USA, but also in several other countries. The shale is penetrated with boreholes that bend horizontally, and then millions of liters of water, chemicals and sand are pumped under high pressure and force open cracks in the rock that release streams of gas. This method is referred to as fracking and is schematically illustrated (see endnotes 27, 31) in Fig. 7.3. Thousands of fracking wells have been drilled in the USA, making this country the world's leading natural gas producer (see endnotes 27, 30, 32). An EASAC statement on shale gas extraction (see endnote 27) contends that there is growing political interest for shale exploration in Europe. The International Energy Agency (see endnote 32) estimates unproven technically recoverable shale gas volumes in Europe to total *13.3 trillion cubic meters*, of which the largest are in Poland and France but also in Romania, Denmark and the Netherlands.

There is, however, plenty of controversy about fracking. Shale gas wells are drilled as part of a cluster with as many as 20 wells or more, requiring large areas of land. The EASAC statement (see endnote 27) points out that besides land, shale gas development requires enormous amounts of water (95% of the fracking fluid), proppants[33] for hydraulic fracturing (~5%) and chemicals (usually less than 1% of the fracking fluid). The enormous quantities of water needed become an issue when existing water supplies are already highly utilized. There are also concerns that the fracturing of shale formations forms fissures which can affect the flows and chemical composition of surrounding aquifers. Furthermore, methane emissions from gas leakage at the wellhead and in the distribution system may be a major factor in populated areas (the global warming potential of methane[34] is about 25 times higher than that of CO_2). In addition, there have been reports of earthquakes induced by fluid injection. Large areas of the USA, long considered geologically stable with little or no detected seismicity, have recently become seismically active. According to McGarr et al.,[35] the increase in

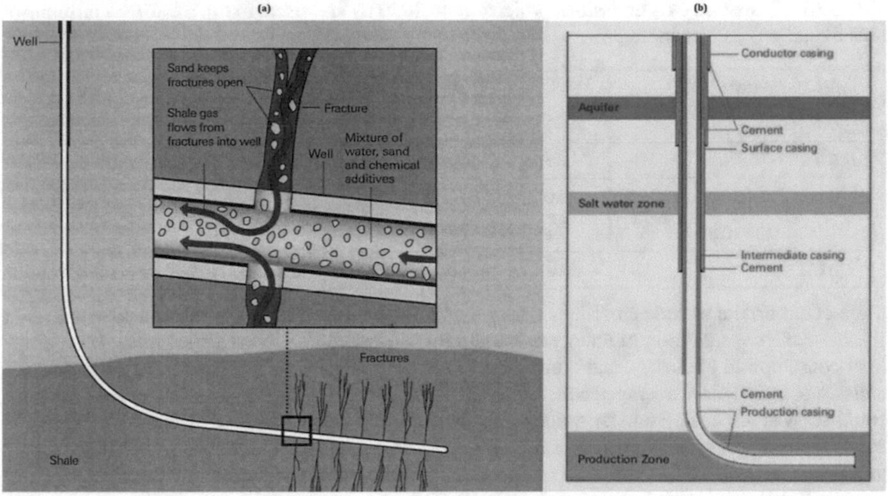

Fig. 7.3 Schematic illustration of fracking (see endnotes 27, 31). (**a**) Fracturing fluids are injected under pressure and cause fractures in the shale. The fractures are propped open by the sand contained in the fracturing fluid; this way the shale gas can flow out of the shale into the well. (Printed with the permission of The Royal Society and The Royal Academy of Engineering) (see endnote 31). (**b**) Cementing the wellbore integrity

earthquake activity in the USA began in the mid-continent starting in 2001 and has continued to rise since; this created activity includes larger earthquakes, several with magnitudes $M > 5$. The authors pointed out that only a fraction of disposal wells has been associated with induced earthquakes large enough to be felt, but the large number of disposal wells contributes significantly to the total seismic hazard, at least in the mid-continent. In some sequences, the magnitudes of the largest induced earthquakes correlate with the volume of the injected fluids (see endnotes 31, 35).

These and possibly other activities show that the age of fossil fuels is far from over. The USA is the world's largest producer of natural gas, and oil is still today the most important form of energy in the world. The increased demand for oil largely originates from its use in the transport sector; emerging economies and economic growth pushes up demand for personal mobility and freight. Most people, certainly those in the developed world, regard mobility a fundamental freedom; yet this freedom increasingly clashes with sustainability, and highlights the problems faced by modern society because of the dependence of modern lifestyles on modern transport systems. Transport is a massive consumer of energy (about 25% of the total energy consumed world-wide) and it has profound adverse impact on the environment and climate change. It should be noted, however, that when we speak of energy for transport we need to define not only the kinds of energy sources but also the kinds of transport we refer to. In Table 7.1 are listed the estimated CO_2 emissions projection to 2050 by end-users in the EU-27 countries.[36] Clearly, the worse means of transportation is road transport.

Table 7.1 Estimated CO_2 emissions projection[a] to 2050 by end-users in the EU-27 (in millions of tons of carbon) (see endnote 36)

End user category	1990	2000	2010	2020	2030	2050
Road transport[b]	695	825	905	980	1002	1018
Rail	29	29	27	27	21	20
Domestic aviation	86	134	179	206	237	244
Inland navigation	21	16	16	17	17	17
Total	**811**	**988**	**1110**	**1213**	**1260**	**1299**

[a]Projections do not include emissions from international aviation and maritime transport
[b]In 2014, the consumption of energy by road transport in the EU-28 countries was 23.2% of the final consumption of energy, and transport was responsible for ~20% of the total greenhouse gas emissions. In 2014, the average contribution of renewable energy sources to transport in the EU-28 countries was 5.9%. These mean values vary from country to country. For instance, in Greece, in 2014, the consumption of energy in transport was 41.6% of the total energy consumption, and transport was responsible for 28% of the total greenhouse gas emissions. The contribution of renewable energy sources in transport was 1.4%

It can be seen from Table 7.1 that the CO_2 emissions from road transport in the EU-member states are predicted to be very high up to 2050, and they are predicted to be very much higher than the emissions from the other three categories of transport combined. In 1990, 86% of the CO_2 emissions came from road transport and this percentage is predicted to remain higher than 80% to 2030. It is thus imperative to reduce greenhouse gas emissions in road transport by the introduction of renewable fuels. The introduction of renewable energy in a mix of fuels for transport in practice means quick production and use of biofuels and use of electrical energy from renewable energy sources for electrical motion (electric cars). Possibly, the biggest challenge facing renewables involves the production of storable fuels (including H_2).

Similarly, significant remains the adoption of new lifestyles and land use patterns, which reduce the need for motorized transport. It is ethically disturbing for the wealthy fraction of the world's population benefiting the most from access to cheap fossil fuels, while the poor fraction being vulnerable to the resulting consequences.

Since carbon sequestration (especially of the diluted CO_2 in transport –exhaust gases) on a scale sufficient to affect the Earth's climate would be an extraordinarily expensive task, emphasis should be placed on alternative fuels for transport. From the perspective of CO_2 emissions (see Fig. 7.2) also, the promotion of greater penetration of natural gas as the preferred energy source choice for transport should be actively pursued. This carbon-based energy source may be considered an appropriate transitional fossil fuel to a low-carbon energy future subject to the qualified assurance that the supply system of this gas is substantially leak-free. The transition to alternative energy sources is difficult but necessary and we should face the challenge: shift from a combustion economy to a solar electric economy and the use of electric cars powered by wind and solar electricity.

7.2.3 *Renewable Energy Sources, Mostly Hydroelectric, Solar, Wind, Biofuels and Geothermal*

Unlike non-renewable energy sources which are exhaustible, renewable energy sources are perceived to be "inexhaustible" for they depend on the Sun, and the Earth has billions of years of sunlight left! No form of energy is however truly "renewable" because energy of a particular form once used is not regenerable. Every energy transformation produces a quantity of low-grade "exhaust" energy that is difficult to reuse. Renewable energy sources while not entirely free of greenhouse gas emissions are nonetheless low-greenhouse-gas production energy sources (see Fig. 7.1).

According to IEA,[37] in 2011 the share of renewables in primary energy use was about 13%, and the share of renewables in the global power mix was ~20%. In the same year, the EU's renewables-based electricity generation was about 21% of the total generation. Accommodation of more electricity from renewable energy sources, often in remote locations, will require additional investment in transmission networks (see later in this chapter). Smil writes (see endnote 7) that "at the beginning of the 21st century, no major economy relied on small-scale, decentralized, renewable energy conversions for anything more than a negligible fraction of its primary energy supply" and that "small-scale, decentralized energy conversions contributed less than 0.5% of the US primary energy supply in 2000". Investing in small-scale renewable energy can help the poor rural regions of developing countries (see, also Sect. 7.4), but they cannot support such basic industries as iron, steel, nitrogen fertilizer synthesis, and cement production. While a clear policy is needed to encourage development of renewable energy sources, public subsidies for these energy sources are usually expensive and run the risk of conveying the impression that renewable energy is cheap, while it is not. Public subsidies in certain countries, for instance Greece, are claimed to have contributed to unacceptable increases in electricity cost rates and to have impacted adversely industrial development.[38] The penetration of renewable energy sources into electricity generation is a goal which should be pursued with vigor, but clearly it cannot be accomplished by perverse subsidies. A realistic policy is thus needed for the totality of renewable energy sources, taking into consideration the new technology and the experience to date.

7.2.3.1 Hydroelectric Power

Hydroelectric power is by far the most significant renewable energy source today (in 2011, it accounted for ~16% of the world's total electricity generation). Most of the underdeveloped potential of hydroelectric power generation is in Africa and Asia; its development on small or large scale will be beneficial to many rural communities. Hydroelectric plants besides energy provide water to communities, but dams may cause environmental problems and population displacements.

The significance of *pumped hydro* for energy storage and recovery must also be stressed. Although there are many energy storage and recovery options which differ dramatically in many ways – for instance, the stored energy density varies from 0.5 kWh/m^3 for pumped hydro to 360 kWh/m^3 for lithium-ion batteries – pumped hydro dominates existing worldwide installed storage capacity of electrical energy. According to a 2010 EPRI report,[39] pumped hydro accounts for 99% of a worldwide storage capacity of 127,000 MW of discharge power.

7.2.3.2 Solar Energy

Sunlight provides by far the largest of all *carbon-neutral* energy sources. More energy from sunlight strikes the Earth's surface every day than all the energy consumed by humanity in 30 years.[40] Presently, solar radiation is exploited through solar photovoltaics, solar thermal electricity, solar heating, and solar-derived fuel from biomass. The efficiency of the solar-to-biomass conversion is very much lower than for the conversion of solar to electrical energy; biomass conversion requires about 100 times more area of fertile land.[41] Solar power plants are also more efficient than wind farms; and, similarly, solar electricity is much more efficient than biofuels. Because photovoltaic cells produce electricity where they are used, they have the potential to reshape the centralized energy economy into something more like the network created by the advent of personal computers, cell phones and the Internet.

Solar energy is highly scalable. Utilization of solar energy on a massive global scale would be extremely desirable, but this would need significant increases in conversion and storage efficiencies.

7.2.3.3 The Role of Light

The energy, frequency, and wavelength of light extend over an incredible range: from 10^{-13} eV, ~50 Hz and ~10^7 m for electromagnetic waves from electrical generators, to >5.7×10^{19} eV, 1.4×10^{33} Hz, and 2.2×10^{-25} m for ultra-high-energy cosmic gamma rays.[42,43]

Light sources can be continuous or pulsed. In the latter case, pulses of light of duration as short or shorter than 10^{-15} s allow study of physical phenomena that occur on this time-scale (even light pulses down to the attosecond scale (10^{-18} s) are now possible and would allow studies of ultrafast processes, for instance, in molecules). This type of research provides fundamental knowledge on the initial stages of the interaction of radiation with matter and the corresponding applications, especially those which depend on the initial short-lived species generated by light at these very short times. The study of the interactions of slow electrons with electronically-excited atoms and molecules generated by short laser pulses provides an exciting new area of research with many novel potential applications such as those (e.g., optical switches) which are dependent on changing the impedance characteristics of matter at the nanosecond and the sub-nanosecond time scale.[44]

The technologies of light (such as the laser light sources, the synchrotron light sources, and the new light sources for illumination) are indeed extraordinary. A few examples of the basic role of light which demonstrate the multitude of the new useful forms of energy based on the transformation of light into other usable forms of energy and their dependence on the photon energy and the nature of the material with which the light interacts are given below.

Solar energy is abundant, environmentally friendly and universally available. Sunlight is by far the largest of all carbon-neutral energy sources. The economic utility of the radiation flux of solar energy, however, is limited by the finite rate at which it can be captured and by the land area that can be dedicated to harness it. We currently transform the energy of solar photons into usable forms of energy directly or indirectly, principally through solar photovoltaics (PV), solar thermal electricity (concentrating solar power), solar heating, and solar-derived fuel from biomass. Many challenges and opportunities for the development of solar energy as a competitive energy source lie ahead and many technical barriers remain for large-scale implementation of solar energy. There is a need to shift from a combustion economy to a solar electric economy and from liquid and gaseous mobility to solar-energy-generated electric mobility; there is also a need to further develop decentralized small-scale solar-energy systems, and even to consider space solar power and to take advantage of the enormous quantities of energy falling as light on the world's deserts.

(i) *Light for lighting.* It is estimated that today about 25% of the total primary energy consumed world-wide annually is spent for generation of visible light. It is thus imperative to improve existing light sources and to develop innovative ways of generating visible light efficiently. Proper materials and innovative technology have allowed and are allowing substantial increases in the efficiency of light sources such as in the case of Light-Emitting Diodes (LEDs). While for incandescent lamps only 4–5% of the consumed energy is transformed into visible light, the efficiency of LEDs reaches 50–80% (see endnote 9).[45,46] Over the 130 years that have elapsed since Thomas Edison discovered the first light bulb, the efficiency of sources of visible light increased by more than a hundredfold making use of new materials and innovative technology. Basic understanding of the mechanisms via which electrical energy is transformed into visible photons of different colors in novel materials (organic and inorganic) will undoubtedly allow visible light sources of higher efficiency in the future.

Materials open up new possibilities in energy conversions whether these involve the conversion of solar radiation to electricity, fuels or heat; or the transformation of (waste) heat to electricity; or the modification of the solar radiation spectrum for greater efficiency in these and other transformations of light. Low cost organic polymers are being developed for PV; nanomaterials are being developed for energy conversion applications (e.g., nanorods) and for energy storage and recovery; superconducting materials are being searched for efficient transmission and distribution of electrical energy.

(ii) *Transformation of solar radiation into other useful forms of energy.* Figure 7.4 shows schematically the transformation of solar photons into electrical energy

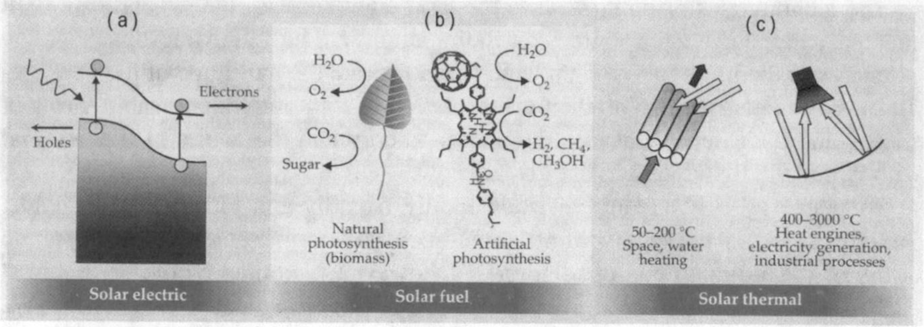

Fig. 7.4 Schematic illustration of the transformation of solar photons into (**a**) electrical energy, (**b**) fuels and (**c**) heat. (Reproduced with the permission of the American Institute of Physics) (see endnote 47)

(via photovoltaics), into chemical energy – fuels (via photosynthesis), and into "usable" heat which is subsequently converted into electrical energy.[47] In Fig. 7.4a the radiant energy from the Sun is converted to electrical energy in two steps. First, solar radiation is absorbed by a suitable material (for instance, a semiconductor) of the PV cell and a portion of it generates electron-hole pairs. Second, the charges (electrons and holes) of the pairs are separated and transported to the respective electrodes creating a current in the external circuit, i.e., electricity. A good portion of the radiant energy ends up as kinetic energy of the electrons and is dissipated in the material, eventually ending up as non-usable heat.

Let us follow a little further the energy conversion process in Fig. 7.4a from the point-of-view of the absorption of the solar radiation in the material. The fraction of the solar energy which is absorbed to the incident energy depends on the energy spectrum of the solar radiation (schematically depicted in Fig. 7.5a) and the absorption band(s) of the solid (schematically depicted by the energy gaps E_g of a single-band semiconductor material to the right, Fig. 7.5c and d). A substantial portion of the absorbed solar energy is wasted as heat through several energy degradation steps and only a small fraction of the absorbed light is converted into electrical energy. Higher energy conversion efficiencies are possible if, for instance, materials are found which modify the solar energy spectrum from that shown in Fig. 7.5a to that shown in Fig. 7.5b so that it matches better the energy gap(s) of the absorbing materials.[48] Suitable materials (converters) could transform photons whose energy $h\nu$ is at least twice the energy gap E_g of the semiconductor ($h\nu \geq 2E_g$) into two photons each of energy $h\nu'$ equal to (or in excess of) E_g which could subsequently be both absorbed, producing two electrons rather than one (Fig. 7.5c). Similarly, in the case of solar photons with energy $h\nu$ less than the energy gap E_g, suitable materials could transform the energy of two such photons into one photon whose energy is higher than the energy gap ($2h\nu = h\nu' \geq E_g$), which could then be absorbed and produce electrons (Fig. 7.5d). It could also be possible to achieve absorption of a larger portion of the solar radiation by using materials with appropriate energy band structure (see endnote 48).[49] This way the efficiency of PV cells could increase considerably.

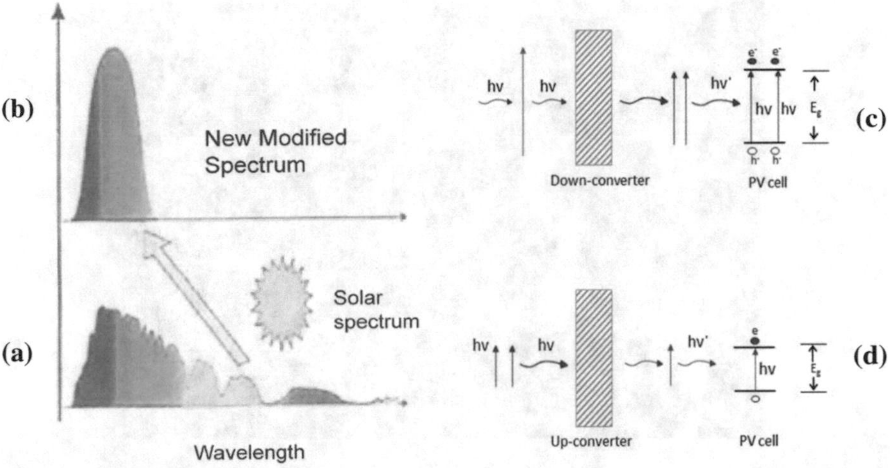

Fig. 7.5 (**a**) and (**b**): Possible shift of the energy distribution of solar photons from (**a**) to (**b**) so that it matches better the energy gap(s) of the absorbing material (see endnote 48). (**c**) and (**d**): Suitable materials (converters) could transform solar photons as shown in figures (**c**) and (**d**) for more efficient conversion of solar radiation into electrical energy (see endnote 49)

In the second example (Fig. 7.4b), that of photosynthesis, radiant energy from the Sun is converted to chemical plants, algae and other species,[50] and in the third example (Fig. 7.4c) sunlight is absorbed by a suitable material and is converted into heat energy, which is subsequently employed for various purposes (e.g., for space or water heating). Alternatively, solar radiation is indirectly converted into electricity via thermal energy production and storage (as in Concentrating Solar Power, CSP). In CSP technologies, solar photons are concentrated and directed to (focused at) a small area of a central receiver system where they heat up a suitable liquid (molten salt) to elevated temperatures; the sun-heated molten salt transfers and stores the heat to a thermal storage system where it can be stored for many hours, even days. Subsequently, as needed, the stored heat is extracted by heating a suitable heat-transfer liquid (e.g., molten salt) which transfers the heat to where it is converted into kinetic energy (steam) and then into electrical energy.[51,52,53]

Figure 7.6 shows one of these types of technologies which transform solar radiation firstly into heat and subsequently into electricity; it is the Solar Tower Power System in Spain. The Solar Tower Power Systems use a solar tower as the central receiver and circular rows of large parabolic mirrors (called heliostats). Each mirror follows the Sun and collects and focuses the sunlight onto the central receiver, which is located at the top of the Tower (in the center of Fig. 7.6). From there, the heat generated is transported and stored as heat in an appropriate energy thermal storage system and at a desired time later it is extracted and is again transported using appropriate liquids (e.g., molten salts) for electricity generation. Other arrangements of Concentrating Solar Power (CSP) are the Parabolic Troughs, the Linear Fresnel Systems and the Parabolic Dishes. Parabolic Trough and Power Tower Systems can provide base load electricity working as hybrids with fossil fuels and using thermal storage. CSP provides dispatchable power and helps match

Fig. 7.6 Gemasolar plant of Torresol Energy in Andalucia, Spain (Torresol Energy) (see endnote 51)

grid supply and demand; it offers a cost-effective way to stabilize the electricity grid and to allow it to accept higher amounts of electricity from stochastic renewable energy sources (PV and especially wind); it can also be used for water desalination and production of energy carriers such as H_2 and syngas. Concentrating Solar Power is important for many areas and countries and for islands. In 2011, there was 1.3 GW of CSP operating world-wide and 2.3 GW under construction (see endnote 51).

7.2.3.4 Energy from Controlled Nuclear Fusion Using Laser Light

Physical science has taught us how to create conditions for the fusion of nuclei of light elements such as those of deuterium (*D*) and tritium (*T*) and produce energy. Although efforts to generate useful energy from controlled nuclear fusion reactions started in 1951, controlled nuclear fusion reactions have not yet become sources of useful energy for humanity. It is believed, however, that this will be achieved one day and it will provide humanity a truly sustainable clean source of energy.

The most critical scientific-technological problem in humanity's efforts to generate useful energy from nuclear fusion is *plasma confinement*. In the stars, gravity pulls the nuclei of the stellar plasma sufficiently close to fuse. This of course cannot be done on Earth. Today, scientific research is focused on two approaches to confine the thermonuclear plasma: *magnetic confinement and inertial confinement.* In the former approach, the plasma is heated with microwave radiation (or other means) and is confined with magnetic fields of high intensity and special shapes (see discussion later in this section) for times longer than ~1 s; efforts in this area are focused on the International Thermonuclear Experimental Reactor (ITER),[54,55] which is expected to be the first fusion device to produce net energy. In the latter approach, very powerful laser pulses are used to confine and heat the plasma. Tiny D-T pellets are hit with high-energy-density laser pulses that rapidly heat the surface of the fusion target, forming a surrounding plasma envelope; the fuel is compressed by the blow off of the hot surface material and is confined for times of ~1 ns. During the final stage of the capsule implosion, the fuel core reaches extremely high densities and ignites at ~100,000,000°C. Thermonuclear burn spreads rapidly through the compressed fuel, yielding many times the input energy.

Laser fusion devices are being developed at the Lawrence Livermore National Laboratory (LLNL) in California and elsewhere. The most advanced facility for conducting fusion experiments employing inertial confinement is the National Ignition Facility (NIF) at the Lawrence Livermore National Laboratory. Figure 7.7 shows fusion with inertial confinement at the NIF[56,57]; the fusion material was D-T pellets 1 mm in diameter and the laser system consisted of 192 powerful neodymium-doped glass lasers (frequency tripled; power 10^{11} kW and laser pulse duration 10^{-9} s). Until today ignition has not been achieved, although in October 2013 a fuel capsule gave off more energy than was applied to it (see endnotes 56, 57). Similarly, in February 2014, it was reported[58,59] that fusion "fuel energy gain" (ratio of energy released by the fuel to energy absorbed by it) exceeding unity (between 1.2 and 1.9) was measured in inertially-confined fusion implosion experiments.

These and other technologies of light are anticipated in the future. Will light lead us to new sources of energy and will photons replace electrons as energy carriers? Will suitable materials be found for the efficient transformation of light into other useful forms of energy? These are open questions, which are crucial for the future of humanity.

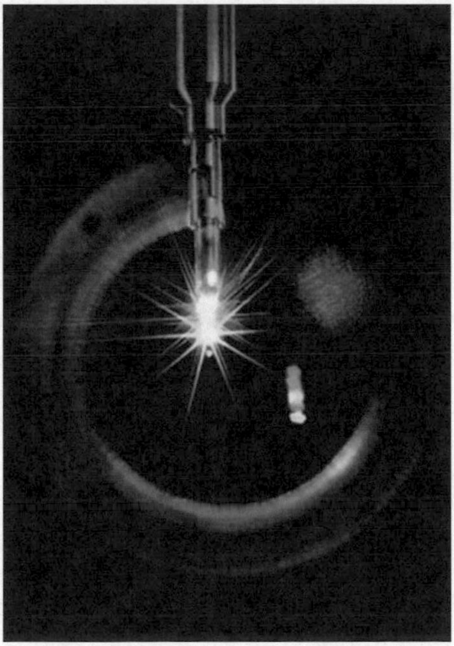

Fig. 7.7 Fusion employing inertial confinement (National Ignition Facility) (see endnote 56)

7.2.3.5 Wind

Wind energy is another secondary product of solar radiation (solar energy converted into kinetic energy of air movement). According to Smil (see endnote 7) the total global installed wind-generating capacity rose from 4.8 GW in 1995 to 17.4 GW in 2000 and to 59.1 GW in 2005; at the end of 2008, it had reached 120.791 GW. While still a tiny fraction (~0.2%) of the primary energy production and a small percentage (~2–3%) of the global electricity generation, experts maintain that the potential of wind power is much greater than these levels.[60,61] Wind-generated electricity is highly stochastic, more stochastic than solar-radiation-generated electricity, and large wind farms have large space demands.

7.2.3.6 Biofuels

Solar radiation is the primary source of free energy for biosynthesis, and biomass is solar energy stored as chemical energy in plants. The solar-to-biomass conversion efficiency is very much smaller than for the conversion of solar to electrical energy; biomass conversion requires about 100-fold more area of fertile land. Biofuels such as ethanol made from corn, sugarcane and cellulose, and biodiesel, methanol and biomass may burn clean, but their production still generates CO_2. Biofuel production might have actually aggravated, rather than eased, greenhouse gas emissions.

Biofuels can be in competition with food and can "distort" global food markets.[62,63] Many argue that biofuels could "starve the poor" and question the wisdom to convert agriculturally productive soil into soil which produces foodstuffs that will be burned as biofuels.

Biofuels derived from plant materials may make society less dependent on oil and other fossil fuels and provide benefit especially to the transport sector. In the future, biofuels will be produced from more complex materials, particularly lignocellulose which is the major component of cell walls and makes up the bulk of the biomass of energy crops such as trees and the perennial grasses. It has also been suggested, that seawater irrigation of desert lands to raise plants adapted for growing in saline conditions may be possible. If indeed this is successful, it could help biofuel production and it could also return a high percentage of fresh water now used for conventional agriculture to other uses. The multifunctionality of forests has also drawn attention in terms of production of biofuels, but there exist many open questions.

7.2.3.7 Geothermal

Geothermal energy is heat from the Earth's interior generated from the radioactive decay of heavy elements; it is a renewable energy source that is available virtually 100% of the time. Either hot dry rock or hot springs could be used for geothermal energy. In half of the world, drilling down to ~2 km reaches rock at ~200°C. According to Goldstein et al.[64] the technical potential for electricity generation can be between 118 EJ/yr (to 3 km depth) and 1109 EJ/yr (to 10 km depth) and for direct thermal uses between 10 and 312 EJ/yr. In 2008, geothermal energy was used for base load electricity generation in 24 countries with an estimated 0.24 EJ/yr of supply. While this is a relatively low value, Goldstein et al. (see endnote 64) estimate that "future geothermal deployment could meet more than 3% of global electricity demand and about 5% of the global demand for heat by 2050". Geothermal energy in combination with other renewable energy sources could help power the world. It must, however, be kept in mind that bringing to the surface of the Earth heat energy from deep under its surface, would add to the heat at the Earth's surface.

7.2.3.8 Other Renewable Energy Sources

Other renewable energy sources include ocean gravitational energy tidal waves, which do not add heat to the planet and can be a source of energy for coastal countries. Waste heat can also be transformed into other uses such as mechanical work, albeit with low efficiency, and solid and liquid waste can be transformed into usable forms of energy. New renewable energy technologies may also be developed, such as space solar where sunlight will be collected in space and the generated power be despatched via satellite stations to the Earth's surface and distributed wherever it is needed.[65] And still other useful forms of energy may be discovered by employing new materials identified by basic research in physics, chemistry, biology and nanoscience, or by exploring exotic ideas.[66,67,68]

7.2.4 *Nuclear Power, from Nuclear Fission and, in the Future, from Nuclear Fusion*

For over half a century, man produces electrical power from the fission of the nucleus of the atom of uranium. When the nucleus of the isotope ${}^{92}U_{235}$ is excited by absorbing a slow neutron, it splits into two nuclei of roughly equal mass each. In the fission process, extra neutrons (n) are produced and a small fraction of the mass of the initial nucleus is converted into kinetic energy of the fission reaction products,

$$ {}^{92}U_{235} + n \rightarrow \text{fission fragments} + \text{neutrons}(n) + \text{energy}(\sim 200\ \text{MeV}). \tag{7.1}$$

In nuclear reactors, some of the neutrons which are produced in reaction (7.1) are used to fission other nuclei ${}^{92}U_{235}$ and maintain, this way, a controlled self-sustained nuclear reaction. A fraction of the neutrons which are generated in reaction (7.1) are absorbed by the non-fissile nuclei of the isotope ${}^{92}U_{238}$ and are transformed into the fissile element ${}^{94}Pu_{239}$, which can be used, unfortunately, to produce nuclear weapons. The fission fragments from ${}^{92}U_{235}$ and other by-products are radioactive and the so-called "nuclear waste" from nuclear reactors contains highly radioactive nuclei, some of which have long half-lives. For instance, the half-life of $^{94}Pu_{239}$ is 24,000 years.

Part of the energy released in fission reactions in thermal reactors is converted into heat which is used to generate electrical energy. The most common type of thermal reactors is the light water nuclear reactor, which uses slightly enriched uranium (3.3–4%) of the isotope $^{92}U_{235}$. The water used in these reactors serves two purposes: as a moderator of epithermal neutrons and as the heat transfer medium. About 80% of the nuclear reactors in operation world-wide for electricity generation are light water reactors (LWRs) and they have a good safety record.[69,70]

In 2001, fifteen (15) countries produced at least 30% of their electrical energy from nuclear fission; over 70% of the electrical energy of France and Lithuania, 34% of Japan, 22% of UK, and 20% of the USA was of nuclear origin (see endnotes 69, 70). However, nuclear energy has not increased as originally expected, although the "burning" of uranium does not generate greenhouse gases. Nonetheless, fission nuclear power continues to be an essential part of the low-carbon electricity generation in the world and is expected to continue to be so for decades. In 2013, nuclear power plants supplied 11% of the world's electricity, down from ~18% in 1996; a total of 30 countries around the world operated 440 nuclear reactors with a gross installed capacity of 392 GW.[71,72] In January 2015, in the EU, 27% of electricity production was obtained from 132 nuclear power plants (see endnote 71).

Fission-generated electric power suffered a setback from three major nuclear reactor accidents: The Three-Mile Island accident in the United Sates (1979), the Chernobyl accident in the former Soviet Union (1986), and the Fukushima accident in Japan (2011).[73,74,75,76] The Fukushima Daiichi nuclear power plant accident raised questions about the future of nuclear power. However, several reports[77] indicate that the Fukushima accident has not so far led to a significant retraction in nuclear power programs in countries outside Europe, except Japan itself. In Europe, changes in nuclear policies have taken place in Germany, Switzerland and Italy, but there seems to have been no changed policies toward nuclear power in countries such as China, India, Russia and South Korea.[78] There is renewed interest in safer, small size (100–400 MW) nuclear reactors. However, the long-term issues of safety, security, proliferation, and fuel-cycle management still haunt nuclear power. Several panels recommend "that countries place their civil nuclear programs under international safeguards run by IAEA (International Atomic Energy Agency), so that spent fuel cannot be diverted for weapons use, and countries that already have nuclear weapons should separate their civil and military programs" (see endnote 78).[79] Similarly stressed, is the need for clearly defined nuclear safety standards to which all countries fully subscribe.

Historically, nuclear power has not achieved broad acceptance by society, and, consequently, nuclear power has not increased as originally envisioned. While society recognizes that every energy source has its own dangers, it considers that nuclear energy is accompanied by serious and unique problems which demand responsible societal action and long-range planning. The main reasons for society's hesitancy to embrace nuclear energy are fears associated with the safety of nuclear reactors, the handling and safe disposal of nuclear waste, the danger of proliferation of nuclear materials and nuclear weapons, and possible long-term health effects from radiation exposure. Society feels uncertain about these issues and is seen unwilling to distinguish between the nuclear reactor as a source of useful energy and the nuclear bomb as a source of unimaginable destruction. Society, thus, remains skeptical as to the advantages of useful energy from nuclear fission, especially when its energy needs have been so far largely satisfied by other sources. It is however argued that nuclear power is needed because it is "clean" and because it provides a good fraction of the world's electricity which is difficult to replace with alternative sources. While nuclear electricity is subsidized heavily by the State, it may become economically more attractive because of taxation on CO_2 emissions. The production of electricity from nuclear fission will most likely continue, and so will probably do its unique problems.

Regarding the long-term issues of *safety, security, proliferation risk and fuel-cycle management*, there is clearly a need for convergence on and adoption of international safety regulations. Several studies and panels have recommended that countries place their civil nuclear programs under international safeguards run by the IAEA, so that spent fuel cannot be diverted for weapons use, and that countries already having nuclear weapons should separate their civil and military programs.

Nuclear waste (a mixture of radionuclides, isotopes that produce ionizing radiation) evokes fear, concern and opposition because society is not convinced that nuclear waste is a manageable problem.[80] An EASAC/JRC report (see endnote 79) discussed and critically assessed the issue of confidence in the indefinite storage of nuclear waste. The issues related to nuclear waste are strongly coupled to the kind of nuclear cycle adopted for treatment of the spent nuclear fuel and its waste.[81] There are basically three options of fuel cycles: the open fuel cycle, the conventional closed fuel cycle, and the advanced closed fuel cycle. The first option, the *open fuel cycle* (or "once through"), only uses part of the energy stored in the fuel. The second option, the conventional *closed fuel cycle*, uses reprocessing of the spent fuel following interim storage. The main components which can be further utilized (uranium and plutonium) are recycled to fuel manufacturing (mixed oxide, MOX, fuel fabrication) and the smaller volume of residual waste in appropriately conditioned form is disposed of in deep geological repositories. The third option, the *advanced closed fuel cycle*, is similar to the conventional one, but the minor actinides are removed during reprocessing. The separated isotopes are transmuted in combination with power generation and only the net reprocessing wastes and those wastes generated during transmutation are, following appropriate encapsulation, disposed of in deep geological repositories. A major factor that determines the overall capacity of a long-term repository is the heat content of nuclear waste.

Recent generations of nuclear reactors improved significantly, reactor safety has been increased, and progress has been made regarding nuclear waste management. While there still remains open the problem of permanent nuclear waste repository, it is claimed that the nuclear waste problems are technologically manageable and technically solvable.

7.2.4.1 Additional Nuclear Options (Breeder Reactors)

Nuclear energy from fission today comes almost exclusively from the fission of uranium and to a much lesser extent from the fission of plutonium and thorium. The source of $^{92}U_{235}$ is limited. However, the sources of uranium are abundant if one considers the possibility of regeneration of nuclear fuel using breeder reactors to burn the complete quantity of uranium, that is, fission reactors with fuel bred from $^{92}U_{238}$. Besides uranium there is also enough energy in the thorium in the Earth's crust to be used in fast neutron reactors.

Fast neutron reactor technology has been developed since the 1950s and several prototype and more advanced reactors were in operation in the world in the 1970s and 1980s. However, since then their operation has been stopped in most countries for technical, economic and political reasons, the main exceptions being Russia, Japan, China and India. Although since the year 2000 there has been renewed interest in the development of fast neutron reactors in several countries, no commercial breeder reactor is operating anywhere in the world today.[82] In the future, breeder reactors might be employed to broaden the availability of nuclear fuel, using natural non-fissionable Th_{232} to generate the fissile daughter isotope U_{233} via the reaction

$$Th_{232} + n \rightarrow U_{233};\ U_{233} + n \rightarrow \text{fission} + 2.3n + \text{energy}(Th\text{cycle}). \qquad (7.2)$$

In reaction (7.2) the non-fissionable natural element Th_{232} is transmuted into the fissile element U_{233} with the absorption of a neutron (n). Subsequently, the daughter element U_{233} fissions with the absorption of a second neutron releasing energy and additional neutrons for the continuation of the reaction (in practice, for every neutron absorbed by the fissile isotope are needed more than two neutrons because of unavoidable losses). The natural element Th_{232} contains only traces of fissile material (Th_{231}), which are not sufficient to trigger the nuclear chain reaction. For this reason, in a nuclear power reactor based on thorium, the natural element Th_{232} would be used in combination with fissile U_{235} or Pu_{239}. Another way to produce complementary neutrons, suggested by Rubbia,[83] is the use of high-energy accelerators in combination with the reactor.

Likely, also, there will be increased efforts to completely burn uranium in breeder reactors via the reaction

$$U_{238} + n \rightarrow Pu_{239};\ Pu_{239} + n \rightarrow \text{fission} + 2.5n + \text{energy}(U\text{cycle}). \qquad (7.3)$$

Here, the natural non-fissionable isotope U_{238} is used to produce the daughter fissile element Pu_{239}.

The breeder reactions (7.2) and (7.3) have considerable advantages[84] in comparison with the thermal reactors based on enriched or natural U_{235}: (i) there is no need for nuclear fuel enrichment since the entire quantity of the natural element (thorium or uranium) is consumed, (ii) the transformation of the entire amount of the natural element into fissile, increases by about 200 times the energy obtained by simply the use of only U_{235}, and (iii) the nuclear waste contains mainly nuclear fragments which are highly radioactive, but have relatively short half-lives. With respect to the concerns of proliferation of nuclear materials which could be used for nuclear bombs, the breeder reaction (7.3) has serious problems since it generates Pu_{239}, while the breeder reaction (7.2) is free of such problems. The thorium cycle has two additional advantages: the thorium reserves are three to four times larger than those of uranium, and thorium has better properties as nuclear fuel than U_{235}.

An interesting comparison by Rubbia[85] of the level of radiotoxicity of nuclear waste from several types of reactors as a function of the time after shutdown, indicates that the level of radiotoxicity of the nuclear waste from breeder reactors based on thorium reaches levels at which the waste may be returned to the environment after about 500 years, while in ordinary PWR (Pressurized Water Reactors) fission reactors presently in use the level of radiotoxicity of nuclear waste remains high for thousands of years.

7.2.4.2 Nuclear Energy from Nuclear Fusion

Efforts to produce energy from controlled fusion reactions began as early as 1951. However, controlled fusion reactions have not yet been achieved as a source of useful energy. The main reason is the extraordinarily difficult physical conditions required for controlled fusion reactions that would yield substantial amounts of useful energy. The generation of electrical energy from nuclear fusion is perhaps the most difficult scientific and technological effort ever undertaken by man.

Scientific data[86] such as those shown in Fig. 7.8 identify the kind of nuclei that can be used as "fuel" in future fusion reactors and the conditions which are necessary for their "ignition".

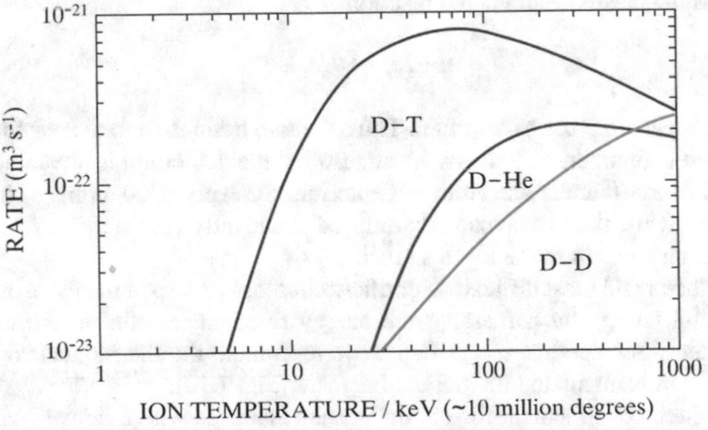

Fig. 7.8 Rate constant as a function of ion temperature in the thermonuclear plasma for *D-T*, *D-He*, and *D-D* reactions (kinetic energy of 1 keV corresponds to ~10 million degrees Celsius) (see endnote 86)

Fusion reactions require light nuclei, such as those of atomic hydrogen (the protons), and the isotopes of hydrogen (the deuterium and the tritium, the former consisting of one proton and one neutron and the latter of one proton and two neutrons) and require for their "burning" temperatures in excess of 100 million degrees Celsius. At these temperatures, the thermonuclear fuel exists in the form of a plasma. Such thermonuclear plasmas must be kept away from the walls of the containing chamber.

The main fuel of a thermonuclear reactor in the future is likely to be deuterium (*D*). The fusion of deuterium (*D*) and tritium (*T*) is especially significant because it has the lowest ignition temperature (about 100 million degrees Celsius) and the largest rate (cross section) for fusion at these relatively "low" temperatures; in addition, it releases large amounts of energy (Fig. 7.9 and Table 7.2).

As can be seen from the data in Fig. 7.9 and Table 7.2, the products of the *D-T* reaction

$$D + T \rightarrow He_4 + n + \text{energy} \tag{7.4}$$

are an α-particle (He_4) with kinetic energy 3.5 MeV and a neutron (*n*) with kinetic energy 14.1 MeV. In comparison with the chemical combustion reaction $2H_2 + O_2 \rightarrow 2H_2O$, the fusion reaction (7.4) releases over 20 million times greater amounts of energy (Table 7.2). The fusion reaction (7.4) is also significant in that the two basic fuel elements (*D* and *T*) are practically inexhaustible.

In a fusion reactor, the kinetic energy of the α-particles will keep the temperature of deuterium and tritium high, sustaining the fusion reaction, while the neutrons (as electrically neutral particles) will escape from the plasma. The escaping neutrons will be slowed down in the medium which surrounds the plasma and the absorbed energy will be transformed into heat and eventually into electrical energy. The slowed-down neutrons will then be trapped in a layer of lithium (*Li*), which surrounds the plasma, and via the reaction

$$n + Li_6 \rightarrow He_4 + T \tag{7.5}$$

they will *regenerate* the tritium fuel. Thus, the two basic thermonuclear fuels, deuterium and tritium, are practically inexhaustible: the deuterium is an abundant element in Nature (there is one atom of deuterium for every 6500 atoms of hydrogen in Nature), while the tritium can be produced abundantly via reaction (7.5) since *Li* is a common element in the Earth's crust.

The crucial and most difficult scientific/technological step in man's effort to generate useful energy by harnessing the energy from nuclear fusion is the plasma confinement. Science has found two ways to confine the thermonuclear plasma: magnetic confinement and inertial confinement (Fig. 7.10).[87,88,89,90,91,92,93,94,95,96]

In magnetic confinement fusion, the plasma is heated with microwave radiation (or other means) and is confined for times longer than ~1 s with magnetic fields of high intensity and special designs (Fig. 7.10). In inertial confinement fusion, a small pellet of material is compressed and heated up with strong laser beams or beams of

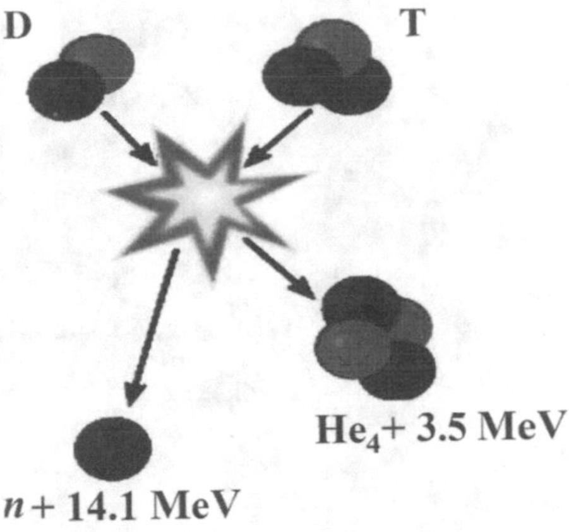

Fig. 7.9 Schematic of the fusion reaction $D + T \rightarrow He_4 + n$ + energy

Table 7.2 The most significant fusion reactions involving deuterium (D)

Fusion reaction[a,b]	Energy[c] (MeV)	Energy[c] (kWh/g)
$D + D \rightarrow T + p$	3.25	22,000
$D + D \rightarrow He_3 + n$	4.0	27,000
$D + T \rightarrow He_4 + n$	**17.6**	**94,000**
$D + He_3 \rightarrow He_4 + p$	18.3	98,000

[a]Tritium (T) is an isotope of hydrogen (H) with one proton and two neutrons. It is radioactive with a half-life of 12.3 years. It does not exist in nature. It is artificially produced, for instance, in collisions of neutrons with atomic lithium (Li).

[b]He_3 is a stable isotope of helium (He_4) with two protons and one neutron.

[c]For comparison, the energy released in the chemical combustion reaction $2H_2 + O_2 \rightarrow 2H_2O$ is ~0.0044 kWh/g.

ions from accelerators. The inertial confinement times are of the order of 1 ns. To produce energy from nuclear fusion, the plasma nuclei must remain confined and their temperature must be kept sufficiently high and long enough to release more energy than that which has been spent to heat them up and to confine them.

Recently, in magnetic confinement experiments, all indicators of plasma quality (plasma density (d), plasma temperature (T), plasma confinement time (τ)), and fusion power (P_f) have increased substantially. As is shown in Fig. 7.11, fusion power levels achieved experimentally between 1975 and 1995 increased by over a factor of 1×10^8 (by over 100 million times) (see endnote 92) (from 0.1 Watt in 1975 to over 10 million Watt in 1995). The value of the so-called triple product, $d\tau E$ (plasma density × plasma confinement time × energy of the plasma nuclei) has reached levels which are ~3–5 times lower than the "breakeven level" (~5×10^{21} m^{-3} keV s) (the value of the triple product for which more energy is generated than it is consumed to heat and to confine the plasma) (see endnotes 93, 95), and about 10 times lower than the "ignition" value (see endnote 91).

Nuclear energy from fusion is considered safe, clean, and inexhaustible. It is relatively free of environmental pollution problems, nuclear waste, and nuclear materials that can be used for nuclear weapons. The effort to develop a useful source

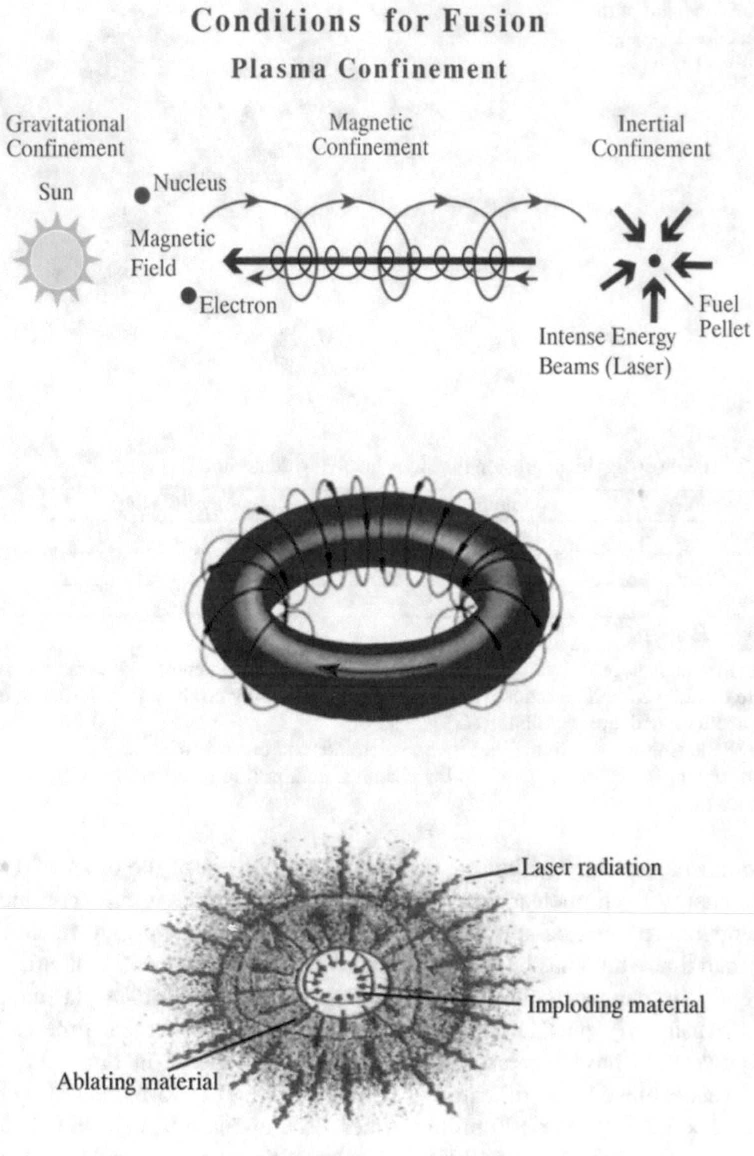

Fig. 7.10 Conditions for fusion: plasma confinement (see endnotes 87–96)

of energy based on nuclear fusion will demand more scientific research and new knowledge (e.g., in the physics of fusion plasma), new technology and materials (e.g., superconducting materials), international collaboration and long-range planning. While substantial progress has been made, useful energy from nuclear fusion is not likely to become available for society to use any time soon.

Undoubtedly, man's efforts to generate on Earth conditions for controlled fusion, will one day be realized. When humanity accomplishes this goal, *the energy from*

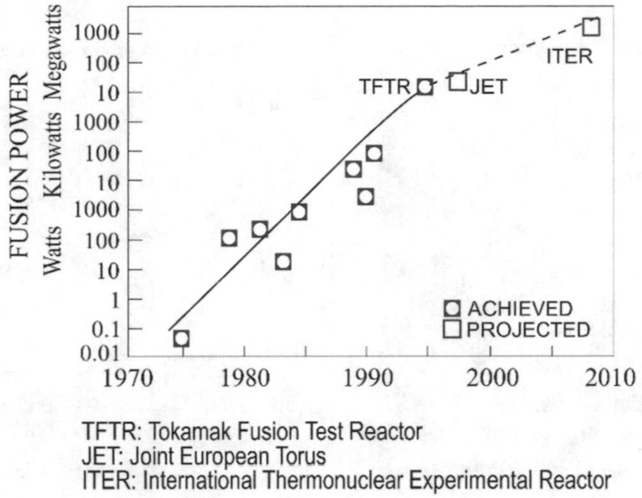

Fig. 7.11 Experimentally achieved fusion power levels between 1975 and 1995. (Data from Tokamak and experiments worldwide) (see endnote 92)

controlled nuclear fusion may well be the main sustainable energy source for humanity. For this kind of fire, as for that from nuclear fission, man must become more responsible.

In the development, use, and management of the nuclear energy sources (from nuclear fission and from nuclear fusion) will depend in large measure man's standard of living and freedoms in the future.

7.2.5 *Trends in Primary Energy Consumption by Fuel and Emerging Primary Energy Mix*

Figure 7.12a shows[97] the current and future trends in primary energy consumption by fuel between 1965 and 2035. It is clear from these data that the total energy consumption will continue to increase. Figure 7.12b gives the emerging primary energy mix per cent of share vs. time (1965–2035). The energy mix of the primary energy sources is clearly changing. The use of oil and coal is declining and that of natural gas is increasing. However, these three carbon fuels (oil, coal and natural gas) will continue to dominate energy consumption for some time.

On the whole, hydro will continue to contribute a large percentage of renewable energy and the gradual transition in the fuel mix toward renewable energy sources[98] will continue to grow. Together with nuclear, hydroelectric power and the other renewable energy sources are expected to account for half of the growth in energy supplies (see endnote 97). Nevertheless, and despite of these changes in the fuel mix and the increase contribution of natural gas, in the decades ahead the data in Fig. 7.12b show that the carbon fuels of oil and coal will continue to dominate.

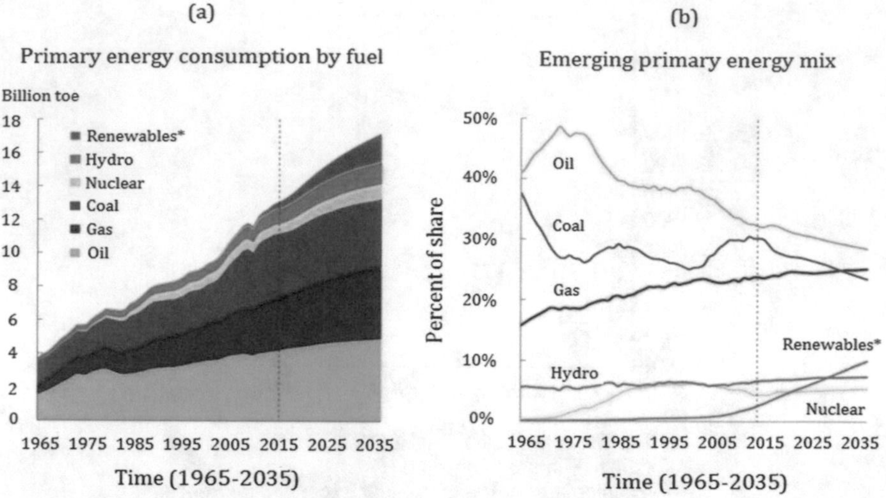

Fig. 7.12 (**a**) Current and future trends in primary energy consumption by fuel between 1965 and 2035; (**b**) Emerging primary energy mix (per cent of share) vs. time (1965–2035) (see endnote 97)

7.2.6 Energy Conservation

Let us look at another side of the energy issue, namely, *energy conservation*. Energy conservation can be seen as the cleanest of all energy sources and as a significant source of energy today and in the future. The best way to conserve energy is, of course, not to use it. Energy conservation is also an effective way to reduce greenhouse gas emissions. Energy conservation and energy efficiency are the most important "technologies" for reduction of CO_2 from the burning of fossil fuels in the decades ahead.

Energy conservation can be looked at as an inexhaustible energy source, the significance of which depends on the responsible behavior of every county and every citizen.[99,100,101] There are countless ways that each one of us can save energy. For instance, reduce energy consumption at home (better home insulation and more efficient heating and cooling, lighting, and electrical appliances; see two examples in Fig. 2.9).

The crucial role of scientific research and technology in the area of energy and the environmental can hardly be overstated. Environmental problems in general require energy for their solution and oftentimes this reduces the energy efficiency of equipment.[102] With research and modern technology, ways are being found to conserve energy, increase energy efficiency and allow compatibility between energy use, the environment and climate change. There are of course numerous other examples of the critical role of scientific research and technology that can be mentioned, especially in the areas of new materials with important energy applications, whether these involve efficient light sources or efficient ways of converting the photons of solar radiation into electrical energy, fuels, or heat.

Besides the technical aspects of the problems of energy and the environment, society's approaches and responses toward energy and the environment are also crucial. We need, for instance, to realize that there are no easy answers to the problems of energy and the environment and that to choose wisely the most efficient and environmentally friendly energy source, or to optimize a combination of them, it is necessary to consider every energy source developed. The fundamental character of energy and the environment, their significance for sustainable development, and their inseparable mutual dependence requires coordinated and sustained action by both government and society alike.

We conclude this section by pointing out four areas where energy savings can be realized:

Efficiency in production, distribution and transport of electricity: The environmental effects are the product of population size, resource consumption levels, behavioral patterns, and inefficient use of technology. Technology and energy resources exist which in various combinations could in time put world energy production and consumption on a more sustainable path. However, they are unlikely to suffice in the absence of steps both to limit world population and to shift consumption patterns towards those that have lower energy- and materials-intensity.

Efficiency and conservation in transport: Most people, certainly those in the developed world, regard mobility a fundamental freedom. Yet, increasingly this freedom collides with sustainability, unless, that is, different types of mobility are introduced. Transport is a massive consumer of energy with profound environmental and climate-change impact, yet modern lifestyles and land use patterns largely depend on the usual motorized transport systems, the most polluting transport systems of all. Gasoline-fueled internal combustion engines can be made considerably more efficient (more than 60 mpg), cars can be made smaller and batteries can be improved to allow putting the electric car on the road.

Efficiency and conservation in lighting: The search for more efficient sources of visible light will continue unabated making use of new materials,[103] as, for instance, in solid-state light sources (e.g., Light Emitting Diodes).[104] An area of significant importance is also light illumination of computer screens. Very important are Liquid Crystal Displays (LCD) which consume less energy and are used for illumination of the screen of laptop computers.

Make waste profitable: Make waste profitable and start by recapturing waste heat and recovery of energy and valuable materials from waste. Heat loss is an inevitable by-product of modern industrial civilization. The conversion of waste heat from vehicle engines or industrial processes into electrical power can be economically feasible if the cost of the conversion device is low, even if maximum efficiency is not achieved, since the thermal energy, which is cost free, would otherwise be uselessly dissipated into the environment. Waste can be used to generate electricity, to recover valuable raw materials and to be turned from a severe problem into a serious resource via a thriving "cyclic economy".[105]

7.3 Electricity

In the future, electricity will become the main energy carrier, more so than it is today, and it will accelerate the transformation of many economies which presently rely on fossil fuels to economies which are based on electrical energy (e.g., electric cars), especially if society is successful in generating electricity from renewable energy sources on a massive scale. There is thus a need to develop and to expand the renewable energy sources, to find economically feasible ways for large scale energy storage and to develop more efficient technology for transport of electrical energy over long distances (see later this section), especially when one considers that renewable electricity and nuclear electricity may be generated far away from where they are consumed.

Electricity is the most important energy carrier today. It is the foundation of the technological civilization and its availability and use underpins all its functions; when energy quality is taken into consideration, the importance of electricity for modern technology becomes even clearer. The diversification of energy sources and their conversion into electrical energy are crucial for the optimum production and use of energy and the conditioning of energy to match the energy needs of modern science-based technology.

In this section, we focus on two important aspects of electrical energy: (i) electrical energy storage and (ii) electrical energy transmission and distribution. The former is especially significant for the optimum and efficient production and utilization of renewable electricity, and the latter for the efficient transmission of electrical energy over long distances.

Electricity generation by fuel and projections by fuel between 1965 and 2035 (Fig. 7.12) show that coal, nuclear, renewables and natural gas will be the dominant energy sources of electricity generation in this time period, with the share of natural gas and renewables increasing. Abundant supply of natural gas spurs greater use of natural gas for electricity generation and transportation and projections show that natural gas will surpass coal as the USA's largest source of energy for electricity generation. Growth in electricity generation from renewables tends to be largely policy- and subsidy-driven.

7.3.1 *Need for Energy Storage*

Electricity is generated from continuous energy sources (sources that are always available, e.g., fossil fuels, nuclear, biomass), from dispersed, intermittent and highly-stochastic renewable energy sources (e.g., wind and solar), and from hydro (flexible generation). The existence of stochastic units in the electricity generation system – and the continuously changing consumption – has major consequences for the stability of the transmission and distribution systems, which are sensitive to voltage and frequency stability. The problems of stability can be manageable when the proportion of the stochastic to the continuous units is relatively small and it is addressed by appropriate changes in the operation of conventional units. Problems of stability appear when the proportionality reaches levels higher from what the existing conventional technologies can manage.

Future power systems would have to be capable of handling problems associated with stochastic, highly distributed, and uncontrollable power sources, and simultaneously capable of handling problems associated with an active and volatile – both spatially and temporally – electricity demand. Controlling such a power system requires changing the prevailing centralized power system control and use instead distributed and stochastic control. Among the challenges for such a change are the *accommodation of the increasing active role of the consumer and the decentralized small-scale energy generation.*

Storage behaves like load when being charged and like generation when being discharged. Storage is needed to decouple generation and load and to balance variable renewable generation by integrating intermittent electricity into the grid system; storage is needed to move electricity through time, providing electricity when and where it is needed (to match the time required for energy generation from renewable sources to the time of demand). Beyond smoothing renewable energy supply, storage can significantly improve the utilization of transmission lines by "metering out" electricity so that the line always runs at maximum capacity. By maintaining constancy of voltage and transmission frequency, storage improves power quality, increases electric grid reliability and asset utilization. Storage offers an alternative to high capital costs of managing grid peak demands and large investments in grid infrastructure. There will, thus, be multiple roles for energy storage and at various levels of the electricity grid: at the generation level, for balancing; at the transmission level, for frequency control and/or investment deferral; at the distribution level, for voltage control and capacity support; at the customer level, for peak shaving and cost management. Electricity which is produced at times of low demand or low generation cost, or from intermittent sources, can be stored as electrical energy or as heat which is then converted to electrical energy and released at times of high demand and high generation cost. Energy storage is a prerequisite for using renewable development in remote locations and for future decentralized energy supply systems. Excess electricity can always be stored cheaply in the form of heat and for a long time.[106]

Interest in power storage (see endnote 39)[107,108] is underpinned by (i) increased deployment of renewable generation; (ii) high capital cost of managing grid peak demands; and (iii) large investments in grid infrastructure for reliability and smart-grid initiatives. Storage's attractive features are: prompt start-up, modularity, easy siting, limited environmental impact, flexibility; electricity storage could simultaneously provide different services in the power system. Some experts maintain that by 2050 the developed world will largely convert to renewable energy and energy storage is an integral part of any renewable energy source system. Since, moreover, the energy mix varies from country to country, the technical and economic optimum concerning the size of an electricity storage system needs to be defined in each case individually and for each energy mix. Energy storage systems can be replaced by conventional energy generation; this, however, can lead to an inefficient use of fossil fuels and demands for investment in additional energy generators with high power output and fast response time.

7.3.1.1 Energy Storage Options

There are many mechanisms for storing and recovering energy and thus many types of energy storage options: electrical, chemical, electrochemical, electromagnetic, mechanical, thermal, pumped hydro and so on (conventional batteries, capacitors, super capacitors, flywheels, compressed air, reversible fuel cells, superconducting magnetic energy storage systems, thermal energy storage systems, etc.). Each energy storage option has unique operational performance, recycling and durability characteristics. Options differ dramatically in many ways; for example, the energy density varies from 0.5 kWh/m^3 for pumped hydro to 360 kWh/m^3 for lithium-ion batteries,[109] a difference by a factor of over 700. Thermal energy storage is especially attractive for solar thermal, because sun-heated molten salt can store energy for hours or even days and can be released when needed to drive a generator.[110]

In the USA, about 2.5% of the total electric power delivered uses energy storage; this is far below the energy storage levels in Europe (10%) and Japan (15%).[111] Existing worldwide installed storage capacity of electrical energy is dominated by pumped hydro, which accounts for 99% of a worldwide storage capacity of 127,000 MW of discharge power. Compressed air storage is a distant second at 440 MW (see endnotes 39, 111). Battery (sodium-sulphur, lead-acid, nickel-cadmium, lithium-ion, and redox-flow) storage is at ~400 MW, but it is recognized that battery systems can offer many high-value opportunities, provided that lower costs can be obtained (see endnotes 39, 107, 108).

An assessment of the economic value of electricity storage on the entire energy-supply system has been indicated (see endnote 108). It is emphasized once again that the exploitation and indeed the dominance of renewable energy sources requires reliable and cost-effective solutions to the problem of synchronizing the production of electrical energy from disperse and discontinuous sources with variable consumption demand. Two critical factors for the success of this goal are: *energy storage and proper (smart) grids for the transmission and distribution of electrical energy.* Large-scale utilization of solar radiation for electricity generation is possible if the effective technology for its storage can be developed with acceptable capital and running costs. If excess renewable electricity cannot be stored, ways must be found to use it where it is produced.

A number of studies have compared power and energy densities for different rechargeable batteries (see endnote 111).[112] Figure 7.13a compares (see endnote 112) the power density as a function of energy density of various energy storage devices. Batteries store more energy per unit weight than electrochemical capacitors. The capacitors, shown on the low-energy-density end, refer to the dielectric and electrolytic type which are widely used in power and consumer electronic circuits. These types of capacitors have very high power, very fast response time,

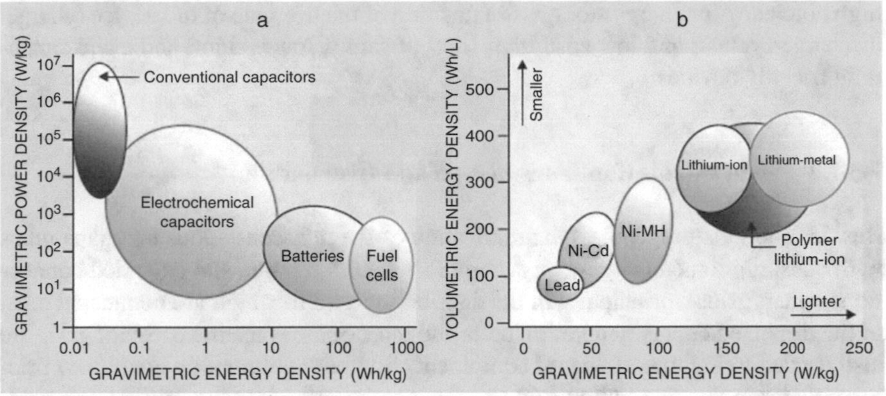

Fig. 7.13 (**a**) Power density as a function of energy density for various energy storage devices. (**b**) Energy per unit volume (Wh/L) as a function of energy per unit weight (Wh/kg) of common commercial batteries. (Reproduced by permission of American Institute of Physics) (see endnote 112)

almost unlimited cycle life, and zero maintenance. However, their energy density is very low (less than ~0.1 Wh/kg in most cases); because they store very tiny amounts of energy, they are not suitable for applications in which significant energy storage is needed. Batteries can store a great deal of energy but have low power – they take a long time to be charged or discharged. Rechargeable lithium batteries are a promising energy storage technology. Batteries based on lithium, store the highest amount of energy per unit of weight (Fig. 7.13b). Rechargeable batteries include those based on lead, nickel-cadmium and nickel-metal hydride. Battery performance is the major barrier to electric vehicles. The most significant issue is energy storage density by both weight and volume. Li-ion batteries for commercial electronics and automobile applications enabled this technology to address reliability, cycle life, safety and other factors that are as important for stationary energy storage. The research environment for developing new low-cost materials is well established and recent efforts directed at low-temperature processing and renewable organic electrodes provide the basis of future advances in the field. However, it is the anticipated production volume for the electric vehicle market that can lead to improvements in the manufacturing process and provide an economy of scale that will bring down the cost.

Today's electrical energy devices – chemical storage (batteries) or electrochemical capacitors – are not capable for meeting tomorrow's energy storage requirements. Greatly improved electrical energy systems are, for instance, needed to progress from today's hybrid electric vehicles to plug-in hybrids to all-electric vehicles. Battery performance is the major barrier to electric vehicles. Batteries to power a car for some 500 miles on a single charge are highly desirable, and battery cost and weight must be brought down and energy and power density must increase.[113]

Material science lies at the heart of innovation in energy storage. Research and development is needed in search for materials with *high-energy density* (Wh/kg) and *high-power density* (W/kg) (high efficiency for recovering the stored energy);

high efficiency for energy storage as a function of the life span of the device (charge/discharge cycles); and low cost, high-level of safety, low weight, and environmentally friendly devices.

7.3.1.2 Smart Grids (for Transmission and Distribution)

Besides energy storage, the synchronization of the non-continuous energy sources with consumption can largely be accomplished with reliable and extended connections (smart grids), principally in the distribution grid to which are connected most of the dispersed energy generation units and most of the consumers. Smart grids are also needed to incorporate smart equipment and house automation about new pricing policies in the distribution grid and the incorporation into the grid electric vehicles and energy conservation measures.[114]

7.3.2 Electrical Energy Transmission and Distribution

The principal technology used today for long distance high power transmission is high-voltage overhead transmission lines (both for alternating current, AC, and for direct current, DC). However, several challenges have limited wider application of this technology. Siting overhead power lines is regarded as one of the most difficult issues facing today's electrical energy transmission projects. There is pubic opposition to siting overhead power lines and the reasons vary from right-of-way requiring broad land corridors, concerns over aesthetics and real estate evaluation issues, to concerns over possible health effects of the electromagnetic fields generated by them.[115] Underground gas-insulated transmission lines, where the high-power transmission lines are located in a pipe and are insulated at room temperature by a high pressure ($\geq$ 4 atm) gaseous dielectric having dielectric strength much higher than that of air (for instance, mixtures of SF_6 in N_2 at a concentration of $\geq$10%[116,117,118]) have been proposed for some time now, but apparently have not had wide usage for *long-distance* electric power transmission.

Three technologies appear to be attractive for long-distance electric power transmission and distribution: Gas-insulated transmission lines, superconductor electricity pipelines, and room-temperature superconductors. The first has not been widely applied for long-distance power transmission but for limited short transmission lines, the second is been suggested for long-distance power transmission, and the third is still to be developed.

Gas-insulated transmission electrical equipment Sulphur hexafluoride (SF_6) has been and is being used by the electric power industry in power transmission and especially distribution (foremost in circuit breakers and gas-insulated substations), because it is a strongly electronegative gas (SF_6 captures very efficiently slow electrons forming negative ions)[119,120] and hence has a very high dielectric strength and excellent power-interruption properties (see endnotes 117, 118).[121] The SF_6 molecule is, however, a potent greenhouse gas (see endnote 34)[122,123]; it stays in the atmosphere for a very long time (halftime ~3200 years) and its global warming potential is ~25,000 greater than that of CO_2 (that is, one molecule of SF_6 causes as much damage in terms of global warming as about 25,000 molecules of CO_2). Basic and applied research (see endnotes 116–121) led to the identification of gaseous media to replace powerful greenhouse gases such as SF_6, which can be used for power transmission and distribution.

Underground gas-insulated transmission lines require only a few feet of permanent right-of-way, are free of possible electromagnetic field affects, and are free of ice storms-, weather- and lightning-problems; they are also free of concerns about aesthetics.

DC superconductor cables Superconductor cables are of two types: *Low-temperature superconducting cables and high-temperature superconductors* (see also Chap. 4). The former is a reality, the latter are yet to follow. Like the gas-insulated transmission lines, low-temperature superconducting cables require only a few feet of permanent right-of-way and are free of the problems faced by the overhead transmission lines mentioned earlier.

The distance between urban centers requiring more electric power and new sources of electricity generation is getting longer, and transmission of large amounts of electric power over long distances is becoming an important challenge. Should large amounts of remote electricity generation become the norm, appropriate technology for transmitting high levels of power over long distances will become necessary. The challenges presented by long-distance high-power transmission lines, can be effectively addressed by DC superconductor underground cables, which, when coupled with voltage source converters, enable multi-terminal transmission over long distances, able to support the connection of many renewable energy "farms" over a wide geographic area.[124,125,126] According to several studies, superconducting cables can carry up to 5 GW of power at 200 kV instead of 750 kV (or higher), which is typically required for conventional long-distance DC lines. The lower voltage is a significant advantage for converting AC to DC and vice- versa. The Electric Power Research Institute (see endnote 125) has shown that high capacity superconducting DC transmission cables can be integrated into the grid with no loss of stability or reliability. Superconductor electricity pipelines have high-power carrying capacity, low power losses, multi-terminal access allowing them to accept power from multiple distributed sources and deliver power to multiple distributed destinations. Superconductor cables are ideally suited to move renewable electricity from where it is generated to where it is needed even for distances over 3000 km. However, low-temperature superconductors require cooling: s*uperconductor materials must be refrigerated to exhibit their ideal electrical characteristics.* The cables are thus cooled cryogenically with conventional liquid nitrogen (LN_2) refrigeration systems that are widely used in industry. While some power is required for refrigeration (this lowers the overall system efficiency), superconductor power cables still have much higher overall efficiency than other long-distance transmission systems (see endnotes 124, 125). What would happen if LN_2 leaks? Nothing; LN_2 will evaporate and it is not toxic (we breath it continuously when we breath air).

High-temperature superconductors The development of high-temperature superconductors will signal the "age of magnetism" and will impact technology profoundly, just as electricity and electromagnetism did in the previous century (see Chap. 4). As mentioned in an overview of the DOE Superconductivity program,[127] "High-temperature superconductivity has the potential for achieving a more fundamental change to electric power technologies than has occurred since the use of electricity became widespread nearly a century ago. In many ways, the transmission and distribution of electricity are poised for advancement the way the Internet was poised before its take off in the 1990s. Just as fiber optics enabled the 'information superhighway' by supplanting lower-capacity copper, superconductivity is enabling an 'energy superhighway' by supplanting copper electrical conductors with a ceramic superconducting alternative that has higher capacity while eliminating resistive losses".

Room-temperature superconductors ($T = 0°C$) Several programs worldwide are aimed at moving high-temperature superconducting power cable systems out of the laboratory and into applications. *Finding a room temperature superconductor will have enormous technological applications in many areas including energy.*

7.4 Energy and Poverty

Energy is a major factor of social wellbeing and the key for poverty eradication. The basic challenge of any society is its moral responsibility to make modern forms of energy, especially electricity, accessible to and affordable by **all** its citizens. Widespread energy poverty was a persistent challenge to society in the previous century and continuous to be so in the present one. It is, thus, apparent that the future of civilization in general, and the eradication of poverty in the poor regions of the world in particular, will depend not only on the total amount of energy humanity will have at its disposal, but perhaps more so on how that energy is distributed among the peoples of the world and is used by them. It is gradually being acknowledged that *access to energy is a human right and a moral responsibility of civilization.*[128,129]

The dimensions of energy poverty are vast and their consequences serious, often devastating. As we have said elsewhere in this book, energy is an input for nearly all basic human needs, goods and services – food, water; shelter, sanitation, healthcare, education; lighting, heating and cooking; transport, communications, information. Access to and affordability of modern forms of energy, especially electricity, is a prerequisite for human development, increased productivity and income.[130,131,132] It has long been known that there is a positive correlation between health and income (life expectancy and income level[133,134]), as well as between electricity access and a variety of quality-of-life indices such as human development index and educational attainment.[135,136,137,138]

Almost every problem the poor countries of the world have, relates directly or indirectly to energy. To wipe out their poverty, we must wipe out their energy poverty *and this requires more energy consumption in the future.*

7.4.1 The Fundamental Role of Electricity

Today there is a clear relationship between the consumption of electrical energy and the GDP of a country – the GDP increases with increasing electrical power consumption of that country (Fig. 7.14).[139,140,141,142,143] The correlation between the consumption of electrical energy and the GDP of a country shown in Fig. 7.14 (upper part) is even more impressive for the USA (Fig. 7.14, lower part) where the data show that the increase in the Gross National Product (GNP) of the USA and in electricity consumption go hand-in-hand. Nowhere is the standard of living rooted in energy more than in the USA.

Consequently, every country today searches for more energy. Societies have come to realize that energy holds together the systems on which they depend; hence, the pressing need to marshal energy in a way that sustains and protects those systems. This is what is demanded by the maintenance and the expansion of modern civilization. First it was the industrial ascendance of the countries of Western Europe, later it was the ascendance of the USA and North America, now is the industrial ascendance of the developing countries of the rest of the world. *Thus, modern civilization continues its divergent need for still more energy.*

The fundamental question, therefore, is how the enormous energy needs of the developing countries can be effectively satisfied, without endangering further the Earth and the sustainability of civilization. This challenge necessitates stable, reliable, safe, economically feasible, and environmentally friendly sources of energy, and responsible management of the world's energy resources.

7.4.2 The Poor Regions of the Earth and their Need for Energy: Today's Reality

It is encouraging that the role of energy as a prerequisite of poverty eradication and a higher standard of living is gradually being widely recognized. Yet energy poverty is hard to define and even harder to measure. Some have defined it in terms of access to modern energy services: affordable access to reliable electricity and clean household and cooking facilities.[144,145,146,147] Others, defined energy poverty in terms of the electrical energy required to satisfy basic human needs: 1 kWh per person per day

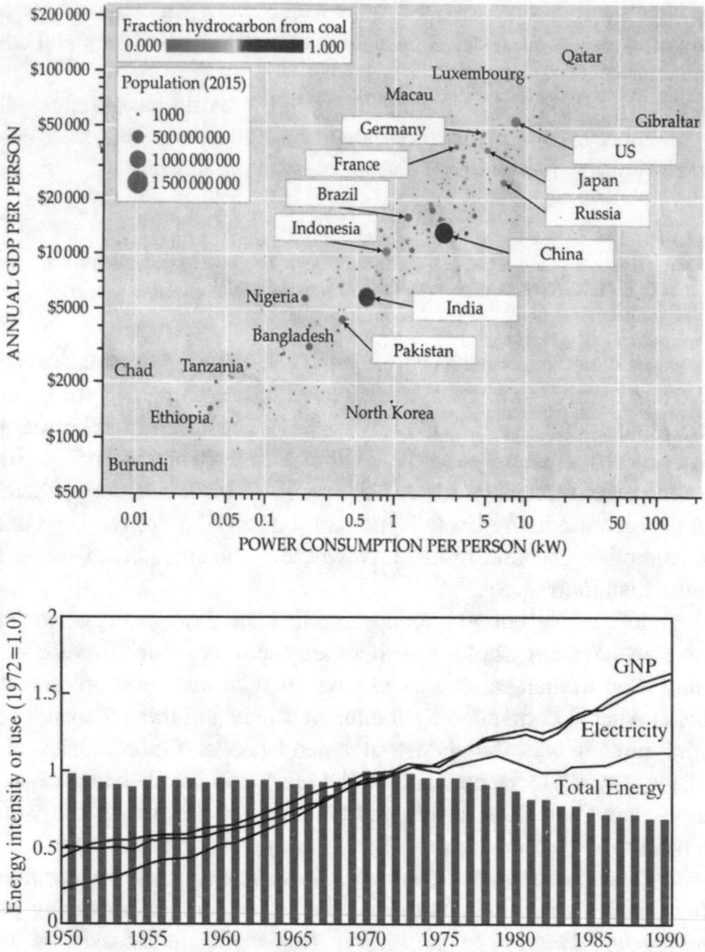

Fig. 7.14 Upper part: Relation between the annual GDP per person and the consumption of electrical power per person for various countries. (From Michael Marder, Tadeusz Patzek and Scott Tinker (see endnote 139), Reproduced with the permission of the American Institute of Physics); see also MacKay (see endnote 140) and Christophorou (see endnote 141). **Lower part:** Energy intensity or use as a function of time between 1950 and 1990 for the USA (Gibbons and Blair (see endnote 142); see also Christophorou (see endnote 143)). The vertical blue strips represent the fraction of the annual consumption of energy and the GNP of the corresponding year (to this ratio was given the value of 1 in 1972)

(500 Wh for fuels and electricity, and 500 Wh for other goods),[148,149] which is close to the ***energy poverty level*** of 250 kWh per household per year for rural households and twice that amount for urban households suggested by IEA[150] and that of Sanchez[151] of 120 kWh per person per year (about 500 kWh per family of four). These threshold energy values have been criticized as inadequate to meet basic human needs (1 kWh per day is equivalent to having four 50-W light bulbs "on" for

5 h a day!). More generally, energy poverty has been referred to as the situation of large numbers of people in developing countries whose wellbeing is negatively affected by very low consumption of energy, use of polluting fuels and excessive time spent in collecting fuel for their basic needs.[152]

Based on such definitions of energy poverty in terms of the electrical energy required to satisfy basic human needs, it is estimated (see endnotes 144, 145, 152)[153,154,155,156,157] that some 1.3–1.6 billion people in the world have no access to electricity and some 2.4–3.0 billion rely on traditional use of biomass for cooking and heating and have incomes less than $2 per day. Most of these people are living in Sub-Saharan Africa and South Asia, mainly in rural areas.

Despite the progress being made, it is argued (see endnotes 154, 157) that, if present trends continue, by the year 2030 some 1.4 billion people will still lack access to electricity and more than 2.6 billion will still rely on traditional biomass fuels, largely because of increasing population in the impoverished areas of the world. For a large fraction of humanity, then, lack of access to modern forms of energy will continue and along with it the lack of provision of clean water, sanitation, healthcare, and economic development. Data indicate (see endnote 154) that while the number of people without access to electricity in the period 1970–2030 will decline in certain regions (especially East Asia, China and North Africa), it will change only slowly for South Asia and it will keep increasing for Sub-Saharan Africa.

An electricity-disconnected world is not just a world without lighting, it is also a lonely world: no TV, no computer, no Internet, no phone, no comfortable reading, and limited access to information. I recall an interesting moment at a meeting of world Academies of Sciences a few years back where a colleague academician from Africa, showing his mobile phone, said: "You see, with this I do not need the expensive phone system which for years you were telling us we could not have because it was too expensive. Now with my cell phone I can communicate directly with anyone anywhere any time." *Of course, but this assumes that he has electricity to charge his phone battery!* If mobile phones are to fulfill the social and economic potential that has so excited development experts, then the capacity for routine recharging is essential.

Contrary to the case of the developed countries, electrical energy is not available (and if available, it is frequently not affordable) in sufficient quantities to the poor regions of the Earth. Yet, unequivocally, income levels correlate with access to modern energy; countries with a large fraction of their population living on an income of less than $2 per day tend to have low electrification rates (see endnotes 156, 157).[158]

In Fig. 7.15 is plotted the GDP (in US $) per capita vs. electricity usage (kWh per capita per year) for EAP (East Asia Pacific) and other countries (see endnote 156). The data in Fig. 7.15 show that large values of the GDP per person go together with high electricity consumption. They also show two other interesting features: (i) a drastic difference in the slops of the low- and the high-income regions, and (ii) a large spread in the data (by a factor of 4–5) of the high-energy consumption region.

Let us then consider the low-electricity-consumption data in Fig. 7.15, say, those below ~2000 kWh per person per year. If we accept 1 kWh per person per day or 365 kWh per person per year as ***the poverty level of energy consumption,*** raising that amount to, say, 1000 kWh per person per year would substantially increase the GDP above the poverty level, while reducing by this amount the energy consumed by the people in the consumption range above 20,000 kWh per person per year would hardly impact their standard of living and, in fact, any such change is within the spread of their incomes. The Earth Institute[159] claims that at the level of electricity usage of ~2000 kWh per person per year, access to modern society needs – modern energy services, more domestic appliances, increased requirements for cooling and heating (space and water), and so on – becomes possible. Nevertheless, IEA (see endnote 150) envisage that the poverty line would rise slowly and reach 800 kWh per household per year by 2030. While such an increase is welcomed, it is still low. By comparison, the average annual household consumption in the 27 EU countries in 2008 was just under 18,000 kWh.

7.4.2.1 What Is it in for Today's Poor?

Although the provision of modern energy services has been broadly recognized as a critical foundation for sustainable development (see endnotes 130, 154),[160,161,162] the UN Millennium Development Goals (MDGs) declared by the UN General Assembly in the year 2000 do not explicitly refer to energy; yet, none of the MDGs can be achieved without the availability of adequate and affordable energy. It is encouraging, however, that a decade after the MDGs were drafted, the basic role of energy in eradicating world poverty has been explicitly recognized by the UN, when the UN General Assembly designated the year 2012 as the "*International Year of Sustainable Energy for All*".[163] It is also encouraging that other international bodies, among them the European Union and the World Bank, are focusing attention on this

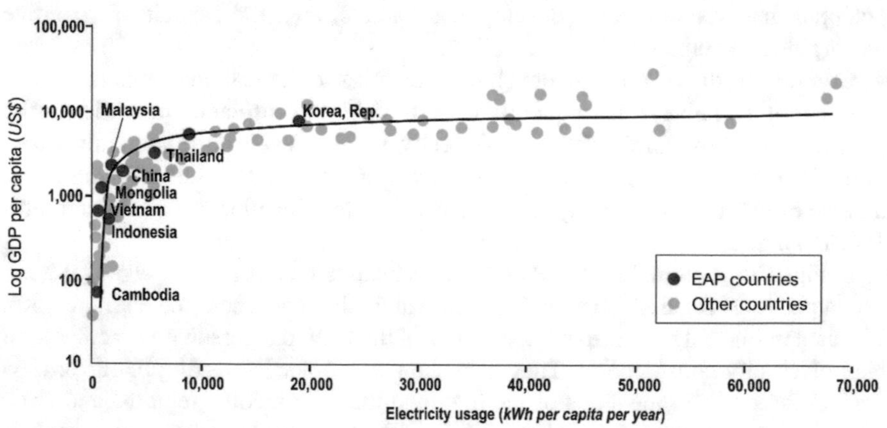

Fig. 7.15 Semi-logarithmic plot of the Gross Domestic Product per person (US$) as a function of electricity usage (kWh per person per year), 2008. (Based on Reference 159)

problem. Calls have also been made[164] for a comprehensive international legal instrument to advance the goal of universal access to modern energy services. Estimates of the cost of programs aiming at universal access to modern energy services by the year 2030 have been made and range between one half and one trillion US dollars.[165]

7.4.2.2 Supply of Electricity to the Energy-Impoverished Regions of the Earth

The supply of electricity to the energy-impoverished regions of the Earth is a key challenge today and in the future. In supplying electricity to the energy-poor areas of the world, the tendency has generally been toward building major fossil-fuel plants located in urban areas and extending existing electricity grids to rural areas, which while sparsely populated carry most of the world's energy-poor peoples. Many argue that this approach and its associated infrastructure can serve well the poor in urban areas, but it is expensive to extend to the poor in rural communities. It is also noted that it is based on the use of fossil fuels – mostly coal – which engenders environmental problems and climate change concerns. Furthermore, long transmission lines present problems of grid maintenance and high-energy losses in electrical energy transmission and distribution[166] (For various approaches to rural electrification see References[131,167]).

Beyond the reach of national grids and independently of them, lie opportunities for energy production in mini-grids and stand-alone off-grid systems from renewable energy sources such as small hydroelectric plants, wind, geothermal, biomass and especially solar. In spite concerns of high starting costs, small-scale renewable energy offers the best opportunity to eradicate energy poverty without adding to greenhouse gas emissions. By all accounts solar electricity will play an important role in improving energy access. Solar power can be installed quickly far off the grid providing enough power for light and basic services. Solar photovoltaics are especially significant as a source of electric power to provide basic services such as lighting, clean drinking water, battery charging and communications.

Figure 7.16 (left) shows the use of photovoltaics at a remote village in Brazil[168]. Here villages never had access to electricity grids, but small photovoltaic systems generate electricity for house lighting. The Smart Village initiative in East and South Africa, and South and Southeast Asia is another international effort to help bring electricity to the poor regions of the Earth[169,170] and serves as a catalyst for development. Figure 7.16 (right) shows Maasai women in Kenya trained in solar panel installation using donkeys to haul their solar wares from home to home in the remote region, giving families their first access to clean and reliable power (see endnote 170).

Small scale renewable energy offers the best opportunity to bring clean energy to the impoverished rural regions of the Earth without adding to greenhouse gas emissions. It is, thus, of utmost importance to increase the efficiency of solar cells and to drive down their cost, and to develop other better and cheaper electricity generating (e.g., wind turbines) and storing technologies to enable large-scale utilization of renewable energy.

Fig. 7.16 **Left**: Small photovoltaics generate electricity for lighting at a remote village in Brazil (see endnote 168). (National Renewable Energy Laboratory, #01270.jpg, "Carimbas" https://images.nrel.gov/bp/#/search?q=Cacimbas&filters=%257B%257D). **Right**: Maasai women in Kenya use donkeys to haul their solar wares from home to home in the remote region. (Smart Villages, Photo by Leopold Obi, 2015) (see endnote 170)

7.5 Energy and Beyond: Conditions for Sustainability of Modern Civilization

Despite the broad recent recognition of the crucial role of energy in the eradication of world poverty, the basic societal challenge still is how to make modern forms of energy accessible to and affordable by *all* peoples and to accomplish this in a world of high and increasing energy consumption, high-energy prices, and large impact of energy consumption on the environment and climate change.

The future will bring increased demands for individual and social responsibility to balance the energy needs of humanity and the environmental consequences of the production and use of energy. Indeed, as we noted earlier in this book, the future of civilization will depend not only on the total amount of energy humanity will have at its disposal, but perhaps more so on how that energy is distributed among the peoples of the world and how it is used by them. The closer we come to the Earth's energy resource limits, the more it will become apparent that *access to sustainable energy will be considered a human right of every person and a moral responsibility of civilization. Man will indeed need wisdom and courage to achieve a precious balance between the restrictions society will impose on him in order to secure for him adequate supply of energy, materials and other resources on the one hand, and, on the other hand, to secure his cultural, civil and human rights, foremost, his freedom, without which there is no civilization at all.*

Modern civilization, then, must change to survive. Two key elements for this change are:

(i) *Effective use of science and technology to meet society's energy needs,* and
(ii) *Guidance from universal human values to secure peaceful coexistence of the peoples of the Earth under conditions of shrinking energy resources.*

Energy will continue to be critical for society. Innovative ways to access known forms of energy and new sources of energy will be sought, and new energy transformations and energy carriers will be searched for.

Energy will continue to be the key in achieving stability of the Earth's climate. Energy production and use will continue to raise fundamental challenges and serious concerns regarding their adverse impact on the environment and climate change. The energy-climate era will thus continue unabated.

Energy will continue to be a major factor of social wellbeing. Ethical questions will continue to be raised about the use of and access to energy. Energy use must be made compatible with human survival, need and dignity.

Electrical energy in particular, will be vigorously sought by every part of humanity because of its significance for economic and industrial development and for every kind of societal infrastructure and social service. A larger fraction of the population of the poor countries of the world will have access to electricity most likely through new and more efficient solar energy technologies. Some envision and optimistically predict that by the year 2050, 95% of the Earth's population will have access to electrical energy.[171]

Further into the future, humanity will probably continue to rely on nuclear energy from fission and, in all likelihood from fusion. Humanity will massively expand solar energy and other renewable energy sources for production of electricity and other energy carriers. The Earth's deserts, many content,[172] may well become humanity's "new oilfields". Humanity may also attempt to "manage" solar energy in space before it reaches the surface of the Earth (see endnote 65), venture into other types of renewable energy[173] growing, for instance, biomass in hitherto unproductive areas using seawater; or produce *methanol through chemical recycling of* CO_2.[174] Or, still, discover new sources of the known forms of energy, or modern technologies to access existing forms of energy, or modern technologies to use energy more efficiently.

Independently, in the far future, if there will be sufficient amounts of energy and if the use of energy expands with proper technology into food production and needed raw materials, humanity could satisfy its needs "using the minerals in the rocks of the Earth's crust, the gases in the atmosphere, and the water of the oceans". However, when humanity reaches this point, its sustainability will depend critically on the functioning of the industrial base of civilization. A serious catastrophe of the industrial base of civilization, would, in all likelihood, render civilization's recovery exceedingly difficult or improbable.[175]

The common future of humanity can be sketched energy-wise this way: A future with different perception of energy where we will learn to consume less energy differently, to travel differently, to produce food differently, to plan our economies differently, to build our cities differently, and to think differently as to the number of people inhabiting the Earth, their interactions and their quality of life. A future where the balance between the energy needs of society and the responsible management of the consequences of the production and use of energy will become multidimensional common responsibility.

References and Notes

1. Earl Cook, *Man, Energy, Society*, W. H. Freeman and Company, San Francisco 1976.
2. www.energy-fundamentals.eu/01.htm.
3. For instance, the idea that heat and work are equivalent was proposed by Julius Robert von Mayer in 1842 and independently by James Prescott Joule in 1843; http://en.wikipedia.org/wiki/mechanical_equivalent_of_heat.
4. Loucas G. Christophorou, *Energy and Civilization*, The Academy of Athens, Athens 2011.
5. Harold J. Morowitz, *Energy Flow in Biology*, Ox Bow Press, Woodbridge, Connecticut 1979.
6. Loucas G. Christophorou, *Place of Science in a World of Values and Facts*, Kluwer Academic / Plenum Publishers, New York 2001, p. 62.
7. Vaclav Smil, *Energy, Myths and Realities: Bringing Science to the Energy Policy Debate*, The AEI Press, Publisher for the American Enterprise Institute, Washington, DC 2010.
8. Jean-Claude Debeir, Jean-Paul Deléage, and Daniel Hémery, *In the Servitude of Power: Energy and Civilization Through the Ages*, Zed Books, London 1991, ISBN: 0 86232 943 4 (translated by John Barzman).
9. Vaclav Smil, *Energy in World History*, Westview Press, Inc., Boulder, CO 1994.
10. Jared Diamond, *Collapse – How Societies Choose to Fail or Succeed*, Penguin Books Ltd., London 2006.
11. Joseph A. Tainter, *The Collapse of Complex Societies*, Cambridge University Press, Cambridge 1988.
12. Todd G. Buchholz, *The Price of Prosperity – Why Rich Nations Fail and How to Renew Them*, HarperCollins Publishers, New York 2016.
13. Keith Chander, *Beyond Civilization*, Authors Choice Press, San Jose, CA 2001.
14. P. B. Weisz, Physics Today, July 2004, p. 47.
15. It is primarily the responsibility of countries whose development and supremacy depend largely on coal (China, India, and USA) to conduct research into “clean” coal burning. Furthermore, since sequestering the diluted CO_2 in transportation exhaust gases is extremely difficult, the world-wide effort to develop “clean” alternative fuels for transport must be accelerated.
16. This corresponds to roughly 2000 W for every person on Earth.
17. Robert F. Service, Science **309**, 22 July 2005, pp. 548–551.
18. IAEA Bulletin, Vienna, Austria, Vol. **42**, No.2, 2000, p. 21; OECD/NEA 2001.
19. J. M. Don MacElroy, Ambio **45**, Suppl. 1, 2016, pp. S5–S14.
20. European Academies Science Advisory Council (EASAC), EASAC Statement, *Greenhouse Gas Footprints of Different Oil Feedstocks*, March 2016.
21. The impact of the burning of fossil fuels on the environment, climate change, and health has been disastrous. The burning of coal, in particular, pollutes the atmosphere with large quantities of carbon-containing compounds (mostly CO_2) and sulphur and nitrogen containing-compounds. The burning of coal

also introduces into the atmosphere particulates, heavy metals (such as mercury and cadmium), radioactive elements (such as radium and radon), heat, and huge quantities of dust.

22. According to Krupp and Horn (see endnote 23), depending on the technology and the type of coal burned, a coal plant emits 1,600 to 2,200 pounds of CO_2 for every MWh of electricity it generates.
23. Fred Krupp and Miriam Horn, *Earth: The Sequel — The Race to Reinvent Energy and Stop Global Warming*, W. W. Norton & Company, New York 2009.
24. Many experts argue that carbon sequestration on a scale sufficient to affect the Earth's climate would be a task of an unprecedented magnitude.
25. James Leigh, *A Geopolitical Tsunami: Beyond Oil in World Civilization Clash*, Energy Bulletin, 1 September 2008 (http://www.energybulletin.net/node/46451).
26. M. King Hubbert, Science **109**, 103–109 (1949).
27. EASAC Statement, *Shale Gas Extraction: Issues of Particular Relevance to the European Union*, October 2014.
28. Michael Marder, Tadeusz Patzek and Scott W. Tinker, Physics Today, July 2016, pp. 47–52.
29. Hydrofracturing technology goes back to the 1940s. However, the assembly of techniques leading to the current production boom are generally attributed to subsequent developments (see endnote 28).
30. Donald Turcotte, Eldridge Moores, and John Rundle, Physics Today, August 2014, pp. 34–39.
31. The Royal Society and Royal Academy of Engineering Report, *Shale Gas Extraction in the UK: A Review of Hydraulic Fracturing*, June 2012.
32. International Energy Agency, World Energy Outlook, June 2013.
33. The most common proppant is sand.
34. Loucas G. Christophorou and Richard J. Van Brunt, National Institute of Standards and Technology, Report NISTIR 5685, July 1995; Intergovernmental Panel on Climate Change (IPCC), the 1994 Report of the Scientific Assessment Working Group of the IPCC.
35. A. McGarr et al., Science **347**, 20 February 2015, pp. 830–831.
36. Joint Research Center (JRC), *Back Casting Approach for Sustainable Mobility*, Luxembourg 2008, EUR 23387/ISSN 1018-5593; *EU Transport GHG: Routes to 2050?*, March 2010.
37. International Energy Agency (IEA), World Energy Outlook 2013 (12 November 2013), Chapter 6 – Renewable Energy Outlook (www.worldenergyoutlook.org/weo2013/).
38. Λουκάς Γ. Χριστοφόρου (Επιμέλεια), *Ενέργεια και Αναπτυξιακός Σχεδιασμός στην Ελλάδα*, Επιτροπή Ενέργειας της Ακαδημίας Αθηνών, Αθήνα 2016. Loucas G. Christophorou (Ed.), *Energy and Development Planning in Greece*, Energy Committee of the Academy of Athens, Athens 2016.
39. Electric Power Research Institute (EPRI), *Electricity Energy Storage Technology Options – A White Paper Primer on Applications, Costs and Benefits*, Palo Alto, CA, 1020676, December 2010.

40. James K. McCusker, Science **293**, 31 August 2001, pp. 1599–1601 (www.nrel.gov).
41. Paul B. Weisz, Physics Today, July 2004, p. 47.
42. Bertram Schwarzschild, Physics Today, May 2010, pp. 15–18.
43. Most of man's vision today lies beyond the light of vision (the energy range of visible light is between ~1.8 eV (~700 nm) and ~3.1 eV (~400 nm)).
44. Loucas G. Christophorou and James K. Olthoff, *Electron Interactions with Excited Atoms and Molecules*, Advances in Atomic and Molecular Physics, Vol. **44**, 2000, pp. 155–292; L. A. Pinnaduwage, L. G. Christophorou, and S. R. Hunter, *Laser-Enhanced Electron Attachment and its Possible Application for Optical Switching*, in Gaseous Dielectrics V, Loucas G. Christophorou and Don W. Bouldin (Eds.), Pergamon Press, New York 1987, pp. 10–17; L. G. Christophorou and S. R. Hunter, *Laser-Activated Diffuse Discharge Switch*, Patent No 4,743,807 (May 10, 1988).
45. Collin J. Humphreys, MRS Bulletin, Vol. **33**, April 2008, pp. 459–470.
46. Λουκάς Γ. Χριστοφόρου, *Υλικά για Ενεργειακές Εφαρμογές*, Επιτροπή Ενέργειας της Ακαδημίας Αθηνών, Ακαδημία Αθηνών, Αθήνα 2010, σελ. 11–22; Loucas G. Christophorou, *Materials for Energy Applications*, Energy Committee of the Academy of Athens, Academy of Athens, Athens 2010, pp. 11–22.
47. George W. Crabtree and Nathan S. Lewis, Physics Today, March 2007, pp. 37–42.
48. Nathan S. Lewis, Science **315**, 9 February 2007, pp.798–801.
49. US Department of Energy (DOE) Office of Basic Energy Science (BES), Basic Research Needs for Solar Energy Utilization, Report of the BES Workshop, 2005. http://www.sc.doe.gov/bes/reports/abstracts.html#SEU.
50. See, for instance, David O. Hall and Krishna Rao, *Photosynthesis*, sixth edition, University Press, Cambridge, UK 1999.
51. European Academies Science Advisory Council (EASAC), *Concentrating Solar Power: Its Potential Contribution to a Sustainable Energy Future*, EASAC Policy Report 16, November 2011 (ISBN: 978-3-8047-2944-5).
52. International Energy Agency (IEA), *Technology Roadmap – Concentrating Solar Power*, 2010.
53. Λουκάς Γ. Χριστοφόρου, *Ελληνικοί Ενεργειακοί Πόροι*, Πρακτικά Ακαδημίας Αθηνών τ. 87 Α΄ σελ. 201–224 Αθήνα 2012. Loucas G. Christophorou, *Energy Resources of Greece*, Proceedings of the Academy of Athens Vol. 87 A΄, Athens 2012, pp. 201–224.
54. B. J. Green, Plasma Physics and Controlled Fusion **45** (2003), pp. 687–706.
55. http://en.wikipedia.org/wiki/ITER.
56. https://en.wikipedia.org/wiki/inertial_confinement_fusion.
57. https://en.wikipedia.org/wiki/National_Ignition_Facility.
58. Philip Ball, *Laser fusion experiment extracts net energy from fuel*, Nature (12 February 2014).
59. O. A. Hurricane et al., Nature **506**, 20 February 2014, pp. 343–348.
60. Ryan Wiser, et al. *Wind Energy*, in IPCC 2011.

61. Gian Andrea Pagnoni and Stephen Roche, *The Renaissance of Renewable Energy*, Cambridge University Press, New York 2015.
62. The Royal Society, *Sustainable Biofuels: Prospects and Challenges*, Policy document 01/08, January 2008 (ISBN: 978 0 85403 662 2).
63. European Academies Science Advisory Council (EASAC), EASAC Policy Report 19, *The Current Status of Biofuels in the European Union, their Environmental Impact and Future Prospects*, December 2012 (ISBN: 978-3-8047-3118-9).
64. Barry Goldstein et al., *Geothermal Energy*, in IPCC 2011. According to IEA (International Energy Agency, *Technology Roadmap – Geothermal Heat and Power*, OECD/IEA 2011) by 2050, geothermal electricity generation per year could reach ~3.5% of global electricity production, and geothermal heat could contribute annually ~3.9% of the projected final energy for heat.
65. The Royal Society, *Geoengineering the Climate: Science, Governance and Uncertainty*, September 2009, ISBN: 978-0-85403-773-5.
66. There can still be exotic ideas of primary energy sources such as quantum vacuum (see endnotes 67, 68).
67. Kimberly K. Smith, *Powering our Future*, Alternative Energy Institute, iUniverse Inc., New York 2005, p. 240.
68. Reference 4, p. 52.
69. Λουκάς Γ. Χριστοφόρου (Επιμέλεια), *Πυρηνική Ενέργεια και Ενεργειακές Ανάγκες της Ελλάδος*, Επιτροπή Ενέργειας της Ακαδημίας Αθηνών, Ακαδημία Αθηνών, Αθήνα 2009. Loucas G. Christophorou (Ed.), *Nuclear Energy and the Energy Needs of Greece*, Energy Committee of the Academy of Athens, Academy of Athens, Athens 2009.
70. Well-run nuclear power plants can operate 95% of the time, typical coal-fired stations 65–75%, hydro 40–60 %, and wind turbines ~25% (Reference 7, p. 40).
71. Akos Horvath and Elisabeth Rachlew, *Nuclear Power in the 21st Century: Challenges and Possibilities*, Ambio 2016, 45 (Suppl. 1), S38–S49.
72. International Energy Agency (IEA), World Energy Outlook 2014.
73. On 11 March 2011, a magnitude-9.0 earthquake on the Richter scale occurred off Japan's northeast coast, triggering a deadly tsunami, which engulfed the Fukushima Daiichi nuclear power station, prompting a series of events that caused huge damage to the three of its six reactors operating at the time and led to a leak of radioactive material.
74. Several reports on these accidents have been published and many lessons have been learned. For instance, after the Fukushima disaster many countries conducted risk and safety assessments of their nuclear power reactors. The Chernobyl and Fukushima accidents demonstrated the need to weigh the risks of evacuation against those of radiation (especially for old people). It has been argued that if iodine was efficiently distributed to the population after the Chernobyl and Fukushima accidents, the thyroid uptake of radioactive iodine would be minimized and similarly the cancer of the thyroid gland. A recent update on the Chernobyl accident (see endnote 75) notes that disaster

preparation is getting more attention since Fukushima. The report states that the Chernobyl nuclear accident remains the worst to date, although debate and controversy surround the extent and the types of damage incurred (from cancers to ecological impact to the effects of massive evacuations on communities and individuals).
75. Toni Feder, Physics Today **69**, April 2016, pp. 24–27.
76. For a history of nuclear accidents see James Mahaffey, *Atomic Accidents — A History of Nuclear Meltdowns and Disasters*, Pegasus books, New York 2014.
77. See, for instance, Science **334**, 21 October 2011, p. 294.
78. World Energy Perspective: *Nuclear Energy One Year After Fukushima*, World Energy Council 2012.
79. European Academies Science Advisory Council (EASAC), M*anagement of Spent Nuclear Fuel and its Waste*, EASAC policy report no. 23, July 2014 (www.easac.eu); Joint Research Center (JRC) (https://ec.europa.eu/jrc).
80. D. Warner North, *A Perspective on Nuclear Waste*, Risk Analysis, Vol. 19, No 4, 1999, pp. 751–758; special issue on *Radioactive Waste* in Physics Today, June 1997; Physics Today, September 2007, pp. 14–16; K. R. Rao, *Radioactive Waste: The Problem and its Management*, Current Science, **81**, No.12, 25 December 2001, pp. 1534–1546.
81. The EASAC report (see endnote 79) discusses several key issues to be taken into consideration in developing national programs for the future management of spent fuel and the waste generated by fuel treatment. Several possibilities exist to deal with spent fuel. Within the so-called "open fuel cycle", spent fuel is disposed of without further use. When "closing the fuel cycle", the energetic component in the spent fuel, plutonium and uranium, is extracted (i.e., reprocessed) for reuse. Consequently, in fully closed cycles, up to 50 to 100 times more energy can potentially be generated from the uranium mined originally. In addition, comprehensive recycling and treatment of the used fuel components by anticipated advanced technologies would leave waste material that decays to low levels of radioactivity in less than 1000 years. All these steps involved additional dedicated facilities and require further R&D.
82. The French completed their full-scale breeder, the 1200 MW Superphenix, in 1986; it operated at full power for less than ten months during the next 11 years and was shut down in 1998 (Reference 7, p. 39).
83. Carlo Rubbia, *Supercritical Thorium Reactors*, in *Energy 2050*, International Symposium on Fossil-Free Energy Options, The Royal Swedish Academy of Sciences, Stockholm, October 19–20, 2009, p. 33.
84. See, for example, http://en.wikipedia.org/wiki/Thorium_fuel_cycle; http://www-pub.iaea.org/MTCD/publications/PDF/TE_1450_web.pdf.
85. Reference 4, p. 108.
86. F. K. McGowan et al., Nuclear Data Tables **A6**, 353 (1969).
87. John Sheffield, *Magnetic Fusion Progress: A History and Review*, Oak Ridge National Laboratory Review, No.4, 1987, pp. 1–18.
88. Journal of Fusion Research **10**, 83 (1991).
89. J. Nuckols, J. Emmett, and L. Wood, Physics Today, August 1973, p. 46.

90. J. H. Nuckolls, Physics Today, September 1982, p. 24.
91. R. R. Parker, Journal of Fusion Research **10**, 83 (1991).
92. J. P. Holdren et al., Journal of Fusion Energy **14**, No.2, 213 (1995).
93. J. Wesson, *Tokamaks*, 2nd edition, Clarendon, Oxford 1997.
94. R. D. Hazeltine and S. C. Prager, Physics Today, July 2002, pp. 30–36.
95. R. C. Wolf, Plasma Physics and Control Fusion **4**, R1 (2003).
96. B. J. Green, Plasma Physics and Controlled Fusion **45**, 687 (2003).
97. BP 2017 Energy Outlook, p. 14.
98. "Renewables" here include wind, solar, geothermal, biomass and biofuels.
99. At an international meeting I have attended in 2016, an energy specialist took the position that energy conservation is unnecessary because it only shifts the burden of saving energy to the "masses" telling them they must conserve energy (that is, to sacrifice), while the big companies and the rich do not have to do something like that! There are those who argue that increasing energy efficiency causes greater consumption which offsets the energy saved ("rebound effect") (see endnote 100). However, others disagree and show that rebounds are too small (see endnote 101).
100. Manuel Frondel and Colin Vance, Nature **494**, 28 February 2013, p. 430.
101. Kenneth Gillingham, Matthew J. Kotchen, David S. Rapson and Gernot Wagner, Nature **493**, 24 January 2013, pp. 475–476.
102. Energy efficiency allows us, in effect, to get more work from a given amount of energy.
103. Επιτροπή Ενέργειας της Ακαδημίας Αθηνών, *Υλικά για Ενεργειακές Εφαρμογές*, Ακαδημία Αθηνών, Αθήνα 2010. Energy Committee of the Academy of Athens, *Materials for Energy Applications*, Academy of Athens, Athens 2010.
104. Dan Charles and others, Science **325**, 14 August 2009, pp. 804–811.
105. Λουκάς Γ. Χριστοφόρου (Επιμέλεια), *Διαχείριση και Ενεργειακή Εκμετάλλευση Αποβλήτων στην Ελλάδα*, Επιτροπή Ενέργειας της Ακαδημίας Αθηνών, Ακαδημία Αθηνών, Αθήνα 2015. Loucas G. Christophorou (Ed.), *Management and Use of Waste for Energy Production in Greece*, Energy Committee of the Academy of Athens, Academy of Athens, Athens 2015.
106. Selection of the most effective and efficient storage solution for a specific grid level and location of an electricity system requires adequate modeling.
107. US Department of Energy, *Demand Response and Energy Storage Integration Study*, March 2016 (http://energy.gov/sites/prod/files/2016/03/f30/DOE-EE-1282.pdf).
108. European Academies Science Advisory Council (EASAC), EASAC Policy Report 33, *Valuing Dedicated Storage in Electricity Grids*, May 2017.
109. George Crabtree and Jim Misewich, APS News, 8 December 2010.
110. Increased capabilities for energy storage for specific needs should be pointed out. For instance, (excess) renewable electricity can be used for desalination and can be considered a form of energy storage. This is especially suited for isolated islands with limited water supply.
111. Bruce Dunn, Haresh Kamath and Jean-Marie Tarascon, Science **334**, 18 November 2011, pp. 928–935.

112. Héctor D. Abruña, Yasuyuki Kiya, and Jay C. Henderson, Physics Today **61**, December 2008, pp. 43–47.
113. Matthew Eisler, Physics Today, September 2016, pp. 31–36.
114. Λουκάς Γ. Χριστοφόρου (Επιμέλεια), *Ανανεώσιμες Πηγές Ενέργειας: Προϋποθέσεις Μαζικής Διείσδυσης στην Ηλεκτροπαραγωγή*, Ακαδημία Αθηνών, Αθήνα 2014. Loucas G. Christophorou (Ed.), *Renewable Energy Sources: Conditions for Massive Penetration into Electricity Generation*, Academy of Athens, Athens 2014.
115. NIEHS Working Group Report on *Assessment of Health Effects from Exposure to Power-Line Frequency Electric and Magnetic Fields*, National Institute of Environmental Health Services, Research Triangle Park, NC, NIH Publication No. 98-3981, August 1998. See also, Δημήτριος Κ. Τσανάκας, στο βιβλίο *Ενέργεια και Περιβάλλον*, Επιτροπή Ενέργειας της Ακαδημίας Αθηνών, Ακαδημία Αθηνών, Αθήνα 2008, σελ. 181–192• D. K. Tsanakas, in *Energy and the Environment*, Energy Committee of the Academy of Athens, Academy of Athens, Athens 2008, pp. 181–192.
116. L. G. Christophorou, *Insulating Gases*, Nuclear Instruments and Methods in Physics Research A **268**, 1988, pp. 424–433.
117. L. G. Christophorou, J. K. Olthoff, and D. S. Green, *Gases for Electrical Insulation and Arc Interruption: Possible Present and Future Alternatives to Pure* SF_6, National Institute of Standards and Technology Technical Report, Note 1425, 1997; L. G. Christophorou, J. K. Olthoff and R. J. Van Brunt, *Sulfur Hexafluoride and the Electric Power Industry*, IEEE Electrical Insulation Magazine, Vol. **13** (No.5), 1997, pp. 20–24.
118. L. G. Christophorou and R. J. Van Brunt, SF_6/N_2 *Mixtures, Basic and HV Insulation Properties*, IEEE Transactions on Dielectrics and Electrical Insulation **2**, 1995 (No.5), pp. 952–1003.
119. Loucas G. Christophorou and James K. Olthoff, *Fundamental Electron Interactions with Plasma* Processing *Gases*, Kluwer Academic/Plenum Publishers, New York 2004.
120. L. G. Christophorou and J. K. Olthoff, Journal of Physical and Chemical Reference Data, Vol. **29** (No.3), 2000, pp. 267–330.
121. Loucas G. Christophorou and collaborators (Eds.), *Gaseous Dielectrics*, Volumes I to X, 1978–2004.
122. Intergovernmental Panel on Climate Change, Fourth Assessment Report, 2007.
123. Although the concentration of SF_6 in the atmosphere is today small, its contribution as a greenhouse gas is additive and practically permanent due to its long half-time (~3200 years) in the atmosphere.
124. American Superconductor, *Superconductor Electricity Pipelines*, A white paper, May 2009. American Superconductor, *DC Superconductor Cables for Long Distance Transmission*, Fall EEI Transmission & Distribution Conference, October 12–14, 2009, Kansas City, Kansas.
125. Steven Eckroad, Electric Power Research Institute (EPRI), *A DC Superconducting Cable*, Workshop on Superconducting Direct Current Electricity Transmission, Houston, TX, January 21–22, 2010.

126. http://phys.org/news/2012-07-superconducting-cables-elecricity-grids.html.
127. https://energy.gov/oe/downloads/superconductivity-program-overview.
128. Λουκάς Γ. Χριστοφόρου, *Ενέργεια και Πολιτισμός*, Πρακτικά της Ακαδημίας Αθηνών, τ. 85 Α΄, Αθήνα 2010, σελ. 205–227. Loucas G. Christophorou, *Energy and Civilization*, Proceedings of the Academy of Athens, Vol. 85 A′, Athens, 2010, pp. 205–227.
129. Reference 4, p. 110.
130. Technical Report of Task Force 2, *Sustainable Energy for All*, April 2012.
131. Benjamin K. Sovacool, *Deploying Off-Grid Technology to Eradicate Energy Poverty*, Science **338**, 5 October 2012, pp. 47–48.
132. Benjamin K. Sovacool, Energy for Sustainable Development **16**, 2012, pp. 272–282.
133. World Bank, *World Development Indicators*, Washington, DC 1999.
134. David E. Bloom and David Canning, *The Health and Wealth of Nations*, Science **287**, 18 February 2000, pp. 1207–1209.
135. It has been argued (see endnote 136) that the GDP must be adjusted so as not to include negative/destructive uses of energy or to include other factors such as "the quality of life". Such indices however are difficult to quantify, and for the scope of the present discussion the use of GDP as it has been traditionally defined suffices.
136. See, for instance, Garry Jacobs and Ivo Šlaus, CADMUS, Volume 1, No.1, October 2010, pp. 53–113.
137. See Benka (see endnote 138) for a relation between Human Development Index (HDI) and annual per capita electricity use. Benka referred to a plot by Alan Pasternak of Lawrence Livermore National Laboratory which shows a rough correlation between electricity consumption and HDI. His analysis showed that the HDI reached a high plateau when a nation's people consumed about 4,000 kWh of electricity per capita annually.
138. Stephen G. Benka, Physics Today **55**, April 2002, pp. 38–39.
139. Michael Marder, Tadeusz Patzek and Scott Tinker, Physics Today, July 2016, pp. 47–52.
140. David J. C. MacKay, *Sustainable Energy – Without the Hot Air*, UIT, Cambridge, UK 2009, p. 231.
141. Loucas G. Christophorou, *Energy and Civilization*, Academy of Athens, Athens 2011, p. 86.
142. John H. Gibbons and Poter D. Blair, Physics Today, July 1991, pp. 20–30.
143. Loucas G. Christophorou, *Energy and Civilization*, Academy of Athens, Athens 2011, p. 87.
144. International Energy Agency (IEA), *Energy Poverty – How to Make Modern Energy Access Universal?* OECD/IEA, September 2010.
145. International Energy Agency (IEA), *Energy for All – Financing Access for the Poor*, World Energy Outlook 2011, OECD/IEA, October 2011.
146. World Energy Outlook 2012, OECD/IEA 2012; Chapter 18, Measuring Progress Towards Energy for All.
147. http://en.wikipedia.org/wiki/Fuel_poverty.

148. Shonali Pachauri and Daniel Spreng, *Energy Use and Energy Access in Relation to Poverty*, Centre for Energy Policy and Economics, Swiss Federal Institutes of Technology, CEPE Working paper Nr. 25, June 2003.
149. Jose Goldemberg, *One Kilowatt per Capita*, Bulletin of the Atomic Scientists, Vol. **46** (No.1) 1990, pp. 13–14.
150. International Energy Agency (IEA), *Energy for All – Financing Access for the Poor*, Special early excerpt of the World Energy Outlook 2011, OECD/IEA, October 2011, pp. 3–48.
151. T. Sanchez, *The Hidden Energy Crisis: How Policies are Failing the World's Poor*, Practical Action Publishing, London 2010.
152. Benjamin K. Sovacool, *Energy & Ethics – Justice and the Global Energy Challenge*, Palgrave Macmillan, New York 2013.
153. Jamal Saghir, *Energy and Poverty: Myths, Links, and Policy Issues*, Energy and Mining Sector Board – The World Bank Group, Energy Working Notes, No. 4, May 2005.
154. Vijay Modi, Susan McDade, Dominique Lallement, Jamal Saghir, *Energy Services for the Millennium Development Goals*, UN Millennium Project and the World Bank, Washington, DC 2005.
155. Gwénaëlle Legros et al. *The Energy Access Situation in Developing Countries: A Review Focusing on the Least Developed Countries and Sub-Saharan Africa*, World Health Organization (WHO) and UNDP, New York, November 2009.
156. D. Ostejic et al., *One Goal, Two Paths: Achieving Universal Access to Modern Energy in East Asia and the Pacific*, The World Bank, Washington, DC 2011 (ISBN: 978-0-8213-8837-2).
157. International Energy Agency, UNDP, United Nations Industrial Development Organization, *Energy Poverty: How to Make Modern Energy Access Universal?* Organization for Economic Cooperation and Development, OECD/IEA Paris, September 2010.
158. Often, too, *unreliable* electricity supply impacts heavily on GDP.
159. The Earth Institute – Columbia University, *Measuring Energy Access: Supporting a Global Target*, March 2010.
160. UN-Energy 2005. *The Energy Challenge for Achieving the Millennium Development Goals*, June 2005.
161. United Nations Development Programme, *Energizing the Millennium Development Goals: A Guide to Energy's Role in Reducing Poverty*, UNDP, August 2005.
162. Technical Report of Task Force 1, *Sustainable Energy for All*, April 2012.
163. UN General Assembly, *International Year of Sustainable Energy for All, 2012*, Report of the Secretary-General, 16 August 2012.
164. Adrian J. Bradbrook and Judith G. Gardam, *Energy and Poverty: A Proposal to Harness International Law to Advance Universal Access to Modern Energy Services*, Netherlands International Law Review **57**, May 2010, pp. 1–28.

165. Chakravarty and Tavoni (Shoidal Chakravarty and Massimo Tavoni, Energy Economics, **40**, 2013, pp. 567–573) estimate that wiping out energy poverty by 2030 would increase the world consumption of energy by about 7% (~20 EJ) and would cause a possible increase in the mean temperature of the Earth's atmosphere by less than 0.13°C.
166. According to a World Bank Report (http://data.worldbank.org/indicator/EG.ELC.LOSS.ZS) transmission and distribution losses in 2009 ranged from 79% in Botswana, 73% in Congo Rep., 40% in Iraq, 23% in Zambia, 20% in Pakistan, to 5% in Austria, Belgium, China, Congo Dem. Rep., Czech Republic, Greece, Japan, Singapore, Slovenia, to 4% in Cyprus, Finland, Germany, Korea Rep., Luxemburg, Malaysia, Netherlands, and even to lower values (3%) in Iceland, Israel, Slovak Republic, and Spain.
167. Alexandra Niez, *Comparative Study on Rural Electrification Policies in Emerging Economies*, Information paper, IEA, March 2010.
168. Samuel F. Baldwin, Physics Today, April 2002, p. 62; https://images.nrel.gov/bp/#/.
169. Brian Heap, *Smart Villages – New Thinking for Off-Grid Communities Worldwide*, Banson, University of Cambridge, UK 2015.
170. John Holmes, *The Smart Villages Initiative: Interim Review of Findings*, Technical Report 5, April 2016.
171. The Royal Swedish Academy of Sciences, Energy 2050 International Symposium on Fossil-Free Energy Options, Stockholm, Sweden, October 19–20, 2009. Λουκάς Γ. Χριστοφόρου, *Ενέργεια και Πολιτισμός*, Πρακτικά της Ακαδημίας Αθηνών, Τόμος 85 Α΄, Αθήνα 2010, σελ. 205–227 (Loucas G. Christophorou, *Energy and Civilization*, Proceedings of the Academy of Athens, Vol. 85 A΄, Athens 2010, pp. 205–227).
172. https://newint.org/features/2015/03/01/desertec-long/.
173. See for instance, Kimberly K. Smith, *Powering our Future*, Alternative Energy Institute, iUniverse, Inc., New York 2005.
174. George A. Olah, *Beyond Oil and Gas: The Methanol Economy*, Angew. Chem., Int. Ed. 2005, **44**, pp. 2636–2639.
175. See, also, Harrison Brown, James Bonner and John Weir, *The Next Hundred Years*, The Viking Press, New York 1963.

Chapter 8
The Future: *QUO VADIS HOMO SAPIENS?*

8.1 Change and Its Challenges

Change is what we cannot stop, and science is the prime facilitator of change. Man's *unique ability to cope with and adapt to change may be enhanced or diminished by whatever we will do to ourselves in the future.* Let us, then, look at some of the changes we discussed in this book and ponder over their expected challenges.

A critical challenge which will continue to confront humanity in the future is how to best fathom "our common future" using science and science-based technology. A basic requirement to accomplish this, is to bring into better equilibrium the power science and science-based technology have given us and the moral and political control of that power. How common is the future of humanity going to be? The so-called "common era" depends on many things besides science and technology. The stability of the common society will crucially depend on good will and mutual trust and on the determination of man to achieve just human conditions and rid the world of utter poverty and fear of war and terrorism. In the future, man is likely to be more independent, more dispersed and more electronically linked. Will this common society remember or care about the things we were and the values we once cherished and have lost to get where we will be? Or, would we diffuse into the ocean of a common indifference? We came down the creeks of primitive cultures, down the rivers of civilizations, into the open seas of unified humanity, wrote Margaret Mead[1] and, ultimately, we are becoming part of the vast ocean of gray homogeneity.[2] Will anyone remember where they came from? The answer, of course, will depend on where we will be headed – *quo vadis Homo sapiens?*

While many critical changes need addressing, let us focus on just three, which can prove to be turning points in human history: (i) critical change in science and society to manage the ethics of the control of man over his own biological evolution and to *uphold the image of man and the respect of his dignity;* (ii) critical change in science and society that would allow their mutual accommodation foremost the accommodation of the scientific and the traditional values of man; and (iii) critical change in science and society that would free man from the fear of war, nuclear war in particular.

L. G. Christophorou, *Emerging Dynamics: Science, Energy, Society and Values*,
https://doi.org/10.1007/978-3-319-90713-0_8

8.1.1 Future Scientific and Technological Frontiers and Their Challenges

We would need to understand complexity in society and in science. Societal complexity will increase and will be accompanied by new problems which will require change in societal behavior and novel approaches to its challenges. We would need to better understand societal complexity and the needs of its maintenance, as well as the impact of societal complexity on the stability of future human institutions and traditional values. Scientific complexity will increase as well and new scientific concepts and constructs will be needed for higher level of abstraction and insight in science and for understanding the large-scale behavior of complex systems, biological organisms included.

The technology of the post-industrial society is essentially information, and information cannot function without energy, hence the search for new, abundant, affordable, easily conditioned and environmentally friendly energy sources will continue to expand and to increase in strength. Similarly, explosive new developments lie ahead in many other areas, for instance, in materials with new properties such as nanomaterials and superconductors, in robotics, molecular genetics and molecular medicine, and in the new branches of science that will undoubtedly spring from those fields and others yet to come.

The powerful new scientific technologies will come with new and profound risks that would challenge both science and society. Consequently, it will be necessary to assess and to justify their risk to harvest their benefit. To deny a promising technology by false perception of risk would be unwise. Worse still would be to attempt to ban publicly-funded research fearing its consequences, for such action would only lead to private and for-profit ventures; it is not just difficult, but often unwise, to define limits beyond which science "may never go". However, it must be recognized that the products of scientific research are not normally owned by scientists, but by governments, for-profit organizations, and industries.

The search for knowledge has always been under stress. Progressively, however, we are experiencing more acutely the consequences of the power of scientific knowledge and this will continue at an even higher level in the future. Sixty or so years ago, scientists suddenly became influential in political life because they knew how to make atomic bombs; now it is the biologist's turn; he "knows", or he is in the process of knowing soon, how to change human nature. Both areas will keep reminding us that as science progresses it enhances our freedom and concomitantly limits our freedom by the burden of responsibility the power of scientific knowledge engenders!

Virtually every major issue confronting society has a science and technology component and demands better scientific understanding by the common citizen, that is, *science literacy*. If science wishes to be embraced by society, and it does, science must embrace society; a free and open society is, in turn, necessary to keep science open and universal. Both the scientist and the nonscientist can profit from a better education regarding the relations of science and society.

Recent crises are increasingly seen as crises of ideas, beliefs, and values which have been the foundation of societies in years past. Many believe that in the future traditional human values will be further weakened as scientific technology and its consequences gain momentum. Where will this weakening of ethical standards and this strengthening of scientific technology take us? Will man's ability to create, using technological means, super intelligence, bring the end of man, as it has been prophesized? We have repeatedly noted in this book that ***the challenge of the future is the ethics of the control of Man over his own biological evolution.***

While society will continue to be amazed by extraordinary new scientific developments, origin questions – been trans-scientific – will most likely continue to be questions *science can ask but cannot answer*. There will thus be calls for science to truly become an integral part of culture and civilization, and for science's actions to be guided by human values. Evidently, then, ***unless future applications of scientific breakthroughs are guided by human values, they are likely to harm humanity.***

The challenge of future scientific and technological frontiers to human values will certainly be profound. Conversely, the impact of human values on science will probably restrain science and limit scientific research in certain areas, unless, that is, technology is kept under control and science is embraced by the entire society and becomes truly universal; this cannot be done through fear or material promise, or through biological modification of *Homo sapiens*. ***There is thus a need for critical change in science and society, which would allow accommodation of their respective value systems***.

8.1.2 Change Our Perception of Resources

Modern civilization's emphasis on development, provision of goods and services, abundant personal choice and prosperity has led to many forms of consumption which strain the finite resources of the Earth. Consumer society is all-pervasive and yearned by the entire world. Capitalism, wrote Ferguson[3], invented the concept of "the worker as a consumer" and now this "beast" has overgrown its demand so much so that there is no way to curtail consumption. Consequently, a ferocious international struggle is under way to secure access to scarce resources, and intense international efforts are at work to manage crucial resources. Unsustainable production and consumption levels are a challenge to sustainable development and affect the poorer nations more than the richer; the former nations lack the resilience of the latter. In the past, societies collapsed when their resources, foremost energy, were overexploited and those societies failed to change their behaviour.

A cultural shift would then be necessary, one based on the understanding and appreciation of the limitations imposed on us by the basic properties of matter and by the *need to change our perception of resources, foremost energy*. Some of the long-term energy issues that will challenge science and society are the security and reliability of energy supply, the affordability of energy, and the effectiveness with which a rapid transformation to a low-carbon energy supply system can be accomplished. A faster shift from combustion economy to solar electric economy is then to be expected.

8.2 Toward a Better Future

8.2.1 From a World of Fear to a World of Hope

For this we must rid the world of the perpetual reign of the absolute terror of war and defeat fear by hope and commitment to humanity. In the twentieth century, the benefits of nuclear, chemical and biological technologies were many and profound, but so were the dangers of nuclear, chemical and biological weapons. In the twenty-first century the danger from those technologies will be even greater as they are refined and made more powerful and precise and become available to small groups of people and even to individuals. In all three categories, knowledge of mass destruction appears to quickly replicate. The perpetual reign of absolute terror that began in the previous century will be intensified in the twenty-first century, targeting man. Humanity will face yet another nightmare appearing on the horizon, this time coming rather easily and quietly from biomedical science and technology (foremost genetic engineering).

The challenge to man therefore is to free, at last, science and science-based technology from the bondage of the machinery of war. Humanity, at last, must free itself from the agony of fear, so it can learn to deal justly with people around the world, and make the frontiers of science and science-based technology serve all peoples. Humanity, at last, must make scientific knowledge available to all. It must move, at last, from imprisoned and proprietary scientific knowledge, toward ***an open science*** for the benefit of humanity; from a science and science-based technology tailored to profit and war to a science and science-based technology dedicated to humanity's common future. No matter how we go about solving our common problems, we will not succeed without the cooperation of science and society and for this we need mutual trust and respect. For science to be truly a liberating source for society, society itself must be a liberating source for science. The stability of scientific society must continue, for without it science will not achieve absolute universality and would not be able to bring about a completely unambiguous and immutable transmission of its culture and help make hope a truly shared universal value.

It has been said that there is no more pervasive quality of living organisms than their adaptation. Indeed, as we have mentioned elsewhere in this book, man has an incredible ability to adapt, but this advantage may lead him to adapt to conditions unbefitting of man. We have even adopted to living with the threat of nuclear war! It's time to move decisively from negative to positive adaptation, from negative to positive conditions of life. Without this change, the indiscriminate growth that presently passes for progress is dangerous and fraudulent. We thus face again the issue of cultural values in science and the necessity of moral choice.

8.2.2 Uphold the Image of Man and Respect His Dignity

Let us recapitulate the views of the image of man held by the four perspectives we discussed earlier in this book: The Hellenic, the Christian, the Western and the Scientific.

The image of man according to the Hellenic perspective In the Hellenic perspective of ancient Greece, *Homo sapiens* is the supreme value *par excellence*. Whether we are unique in the cosmos or not, we stand in awe contemplating the way the universe has evolved and a small planet became our home. We are animals, but we are "not nothing but animals". Of the fifty or so million species on Earth,[4] only man has developed civilization and science; only he is adorned by the reciprocal nature of grace, friendship, and love; only he has the gift of free inquiry; and only his actions are subject to moral judgment. Beyond philosophers, poets, artists, mystics, and scientists lie the imposing power of a loving person and the suffering of the pain of conscience.

As the late Professor Constantinos Despotopoulos writes,[5] the Greek poets, philosophers and the seven wise men of ancient Greece, bequeathed humanity examples of life and commands of classical ethics where man is the central value, a "hero and a tragic figure of fate". To Aristotle, man is the supreme value in the world *par excellence* (το «ἄριστον τῶν ἐν τῷ κόσμῳ »), to Protagoras man is the measure of all things («πάντων χρημάτων μέτρον ἐστὶν ἄνθρωπος»), to Democritus man is a "microcosmos", and to Epiktetus man is "a fragment of God". He is urged to know himself («γνῶθι σαὐτόν»), to avoid excesses («Μηδέν ‘άγαν»), to do everything in moderation («μέτρον ἄριστον»), to strive for excellence, heroism, and pure knowledge, for man, according to Aristotle, "by nature wants to know" («Πάντες ἄνθρωποι τοῦ εἰδέναι ὀρέγονται φύσει»). Man is reminded, too, to be conscious of his transient existence, for according to Pindar (Πίνδαρος) «σκιᾶς ὄναρ ἄνθρωπος».

The image of man according to the Christian perspective Deep in the roots of many religions, but especially in the Judeo-Christian belief, lies the uniqueness of man in the cosmos and the supreme trust placed in him by the whole creation. Not only human beings are placed at the center of the universe, but they are considered to be the very purpose of the universe; God and Nature went into a lot of trouble to guarantee man's existence! "What is man, that thou art mindful of him?" asked the Psalmist[6]; and again[7] "For thou hast crowned him with glory and honor".

In Christianity, the image of man is supreme. This preeminence and superior standing of man, and this position of centrality of man in Nature, is because man was created in the image of God[8] and is endowed with immortality. This is the image of man upheld by many past civilizations and this is the image of man still being upheld by billions of people today, scientists and non-scientists alike; its appreciation requires faith, but only "where faith exists can science exist with faith".[9]

According to Eastern Christian Theology[10], man is a person and personhood belongs to every human being because of its singularly unique relation to God. The unique relationship of each human being to God is, thus, what constitutes the uniqueness of a human person, not its individual physical and biological nature. A person can be viewed in the physical and biological manner suggested by science, and in the personal manner suggested by faith. The diminution of man, is, thus, first and foremost a consequence of stripping man of personhood. ***To be human is to be a person.*** Throughout recorded history *Homo sapiens* has been and is a religious person.

The image of man according to the Western perspective Western civilization, writes Ferguson[11], is "the single most important historical phenomenon of the second half of the second millennium after Christ." No other civilization had ever achieved such dominance as the Western over the Rest. Possibly the two most distinct characteristics of Western civilization that are the major reasons for its dominance are Christianity and science.[12, 13] Deep in the foundation of the traditions of the Western civilization lies the concept of man according to Hellenism and Christianity and the clear distinction between man and the animal world: *Man, is the source of all knowledge*. Similarly, the development of science, the application of scientific knowledge to human wellbeing, and the direct and indirect changes brought about by science and science-based technology on man, society, and the environment were until recently uniquely Western; the scientific revolution has been predominantly "Eurocentric".

The recent dominance of the tradition of the Enlightenment, many argue, is intent upon imposing itself around the world and upon eliminating the classical and Christian roots from the Western tradition. Instead of Western civilization, there is supposed to be a global civilization, multicultural and transnational. While historians usually date the beginning of modern era at the end of the fifteenth century, the so-called post-modern era seems to have begun at about the end of the twentieth century.

Today, there exists in the contemporary West no coherent consensus view, such as prevailed in thirteenth century Europe, or in traditional societies still. What we have today, many observe, is a miscellany of notions as to who we are. What is the nature of man? To this question, the materialist and the man of faith give radically different answers. Many in the West today see a diminution of man and an erosion of his image. Extreme world-wide poverty and suffering, horrors and fears of human actions, diminution of the truth in man, are, for many, signs of the degradation of man's image as the supreme value prescribed by old cultures. Modern man is blamed for everything and is given credit for nothing! Independently, w*e seem to build scientific structures and we seem to demolish the image of man who made those very scientific structures possible;* and we seem to have forgotten that science itself is a human activity, not the action of the beast.

The scientific perspective The impact of recent scientific discovery on the image of man is profound and it is negative: Man, is just a part of the animal continuum, nothing different from the rest of the animals; he is neither the measure of all things, nor the center of the universe, nor the source of all value, nor the culmination of terrestrial evolution. Many scientists maintain either that there is no such thing as "human nature" or that altering it is not ethically problematic, which prompts other scientists to ask, "by what standards and on whose authority?".[14]

Others, armed with biotechnology, biomedicine and genetic engineering, prophesize that we are "*en route* to a new stage of evolution, to the creation of the post-human society, based on science and built by technology", thus, debasing humanity as a value. Will intelligently-directed human evolution induce such vast and fast changes in man that humanity may no longer persist as a single species in a few centuries? Clearly, such notions are negative toward man's image. *The ultimate future challenge of civilization, then, is the protection of humanity and the respect of human dignity.*

Man, not science and technology or the rest of life, is accountable for the choices that either enrich or diminish the quality of life. Will future scientific technology make man freer or must he guard against science and technology if he is to remain a free human being, a person? This kind of question will be raised more often in the future, for the opinion advanced by many today is that technological progress erodes the idea of man and threatens to debase humanity as a value. A good part of society sees the need for moral principles to live by, but seriously doubts that science can handle such a role as supreme judge and master of society. They fear that the scientific control of society is real and they reject scientific materialism and reductionism as the one true account of human life. ***And they demand protection of humanity and loyalty to Man***. This demand may become critical for the relations of science and society in the decades and centuries ahead.

8.2.3 Society: From Conflict to Complementarity

The need for the common and the complementary The principle of complementarity was introduced in Physics by Niels Bohr in 1927 to rationalize the wave and particle properties of the electron (and other atomic and subatomic particles).[15] The principle recognizes that it is possible to grasp one and the same phenomenon or aspect of reality by two distinct, mutually exclusive modes of interpretation; by two concepts which cannot logically coexist, but which are both necessary for a more complete description of reality. The two conceptual ways of describing reality complement each other in an almost paradoxical way for neither is comprehended in, or is reduced to, the other. The description of subatomic particles sometimes as particles and sometimes as waves is a classic example of an oxymoron, and yet acceptable to science for it best describes reality via the two concepts and in fact, depending on which aspects of the subatomic particle (of reality) we wish to explore, we can adopt one or the other conceptual method or approach (see endnote 15).[16,17]

The essence of complementarity embodies the truth that reality need be looked at from different angles and from the vantage of different perspectives to achieve a fuller description and comprehension. Complementarity conveys the fact that the nature of reality transcends the capability of any single conceptual scheme or method used for its description and embodies a limitation of our epistemology.

The principle of complementarity has a wider philosophical and societal significance. There exist many aspects of, and relations in, life which are complementary – not opposing – and there is a need to recognize these as such and the existence of complementary approaches to many of society's problems. Often, seemingly conflicting propositions are but complementary. We must then learn to move from conflict to complementarity as we struggle to go from exclusion to inclusion, from conflict to accommodation. A stark example of this is the treatment of human microcultural perspectives as complementary, each illuminating the human culture by complementary views and perspectives. We thus need to broaden the concept of complementarity to include besides philosophical and scientific issues, social and societal aspects and problems and to seek accommodation of differing views or prescriptions of these situations via complementarity when unity evades.

More generally, we see the need to consider that there might be situations where more than two aspects of a given situation or approach can be complementary. Physical reality is what it is; however, our understanding and knowledge of it differs depending on the way we envision it, on how we look at it, and on what method we employ to study it and on the method's limitations. We see in part and in specific ways, not holistically. The whole is perhaps beyond our grasp! And if, additionally, we consider the fact that "reality" is continuously changing, we realize that even our knowledge about the things we know, becomes a function of time, it depends on when we interrogated Nature. Further still, since no observation is without an observer and since the observer is conditioned by his own self, every understanding however obtained has an uncertainty, a personal view of reality.

Science and values enhance our ability to recognize the importance of complementarity and help us curb conflict by accepting seemingly conflictful propositions as complementary.

8.3 A Hopeful Future Based on Science and Values

We live in the age of science and deep down we realize that science is not enough. Beyond science exist the world of philosophy, art, the spiritual and the sacred, and the personal knowledge which includes love and affection and duty and brotherhood and honor. All these are complementary elements of human knowledge and existence; their synthesis defines humanity and what it is to be a person. Science and technology can blur, but they cannot erase man's ethics and values, and scientific and technological culture cannot make the existential and moral questions irrelevant – such questions are being posed anew, emphatically.

It is, then, the mutual responsibility of scientists and society to: (i) curb the power of science to suppress and to destruct and to deploy scientists in the process, (ii) predict, prevent and manage the risk against the idea of Man associated with the advancement of science, and (iii) require the application of scientific knowledge to be compatible with the values of society.

In this book, a case has been made for a hopeful future based on science and values This future will require both science and values to attain a universal status in society. More than in years past, there will be heightened demands for an open mind and a value system merging toward the common and the complementary. Our common knowledge and our common values will enable us to face our common problems and our common aspirations; it will unlock a future of hope, conducive to making the cultural changes we need to live free in a society that would increasingly put visible and invisible controls upon our behavior to secure our way of life. In this common future, we must secure conditions for freedom in science and in society so necessary for both to rid the world of ignorance, hunger and oppression, and to make possible for man to fulfill his obligation to humanity, the rest of life and the Earth. The challenge is ours: ***A science guided in its applications by human values and a value system cognizant of the facts of science and willing to accommodate them.***

References and Notes

1. Margaret Mead, *Culture and Commitment, A Study of the Generation Gap*, The American Museum of Natural History, Natural History Press, Garden City, New York 1970.
2. Λουκάς Γ. Χριστοφόρου, *Αέναη και Κρίσιμη Αλλαγή*, Πρακτικά της Ακαδημίας Αθηνών, τ. 84 Α΄, Αθήνα 2009 σελ. 109-132. Loucas G. Christophorou, *Perennial and Critical Change,* Proceedings of the Academy of Athens, Vol. 84 A΄, Athens 2009, pp.109-132.
3. Niall Ferguson, *Civilization – The West and the Rest*, The Penguin Press, New York 2011.
4. Ernst Mayr, *What Evolution is*, Basic Books, New York 2001.
5. Constantinos Despotopoulos, in *Universal Values*, edited by Loucas G. Christophorou and George Contopoulos, Academy of Athens, Athens 2004, pp. 26-45.
6. Psalm 8, v 4.
7. Psalm 8, v 5.
8. St. Paul, Col. 1:15.
9. Loucas G. Christophorou, *Place of Science in a World of Values and Facts*, Kluwer Academic / Plenum Press, New York 2001, p. 247.
10. Vladimir Lossky, *In the Image and Likeness of God*, St. Vladimir's Seminary Press, Crestwood, New York 2001.
11. Reference 3, p. 8.
12. According to Kurth (see endnote 13) it is widely understood by scholars that Western civilization was formed from three distinct traditions: the classical culture of Greece and Rome, the Christian tradition, and the Enlightenment of the modern era.
13. James Kurth, *Western Civilization, Our Tradition*, The Intercollegiate Review-Fall 2003 / Spring 2004, pp. 5-13.

14. L. R. Kass, *Beyond Therapy, Biotechnology and the Pursuit of Happiness, A Report of the President's Council on Bioethics*, Dana Press, New York 2003.
15. See, for instance, H. J. Folse, *The Philosophy of Niels Bohr*, North-Holland, Amsterdam, 1985; A. Pais, *Niels Bohr's Times*, Clarendon Press, Oxford 1991.
16. For example, when the kinetic energy of a free electron is very small – say, thermal or near-thermal energy – the electron can interact with matter with probabilities (cross sections) that are about equal to those calculated assuming the electron interacts as a wave and not as a particle. For instance, the attachment cross section of a thermal electron (T~300K) to the SF_6 molecule (and to other electronegative polyatomic molecules) has been measured (see endnote 17) and found to be roughly equal to the theoretical cross section determined considering the electron's de Broglie wavelength.
17. L. G. Christophorou, D. L. McCorkle, and A. A. Christodoulides, *Electron Attachment Processes,* in L. G. Christophorou (Ed.), *Electron-Molecule Interactions and their Applications,* Academic Press, Inc., New York 1984, Vol. 1, pp. 477-617.

Appendix: Energy: Scientific, Philosophical and Theological Dimension[1]

Introduction

Aristotle (384–322 BC) was the first to introduce the word *energy* (*ενέργεια*) and the first to comment on its fundamental significance in understanding the physical world. Prior to discussing aspects of the Aristotelian philosophy on energy in relation to current scientific knowledge, reference will be made to the significance of the *concepts* upon which the laws of physics are based, because this is necessary to understand the physical world and the philosophical dimension of energy.

The *physical world* we live in includes all that exists: all space, all material objects, and all non-material reality (e.g., *physical fields* and *physical forces*) that is not visible but can be detected via its interactions with and effects on matter and can be determined and quantified through the means and instruments of science. The *physical world* we live in contains all forms of energy and all transformations of energy; it is the current phase of a long evolutionary course of 13.8 billon years. Our knowledge of the physical world, through the laws of physics and the concepts on which these laws are based, is supremely impressive although it is always emerging. The laws of physics can be changed or amended and the physical concepts that underpin them are being reformed or replaced by others according to the requirements of new scientific knowledge.

Every interpretation and every logical explanation of the physical phenomena, and, consequently, every level of understanding of the physical world based on the physical law, depends directly or indirectly on the physical concepts that science uses at that time. Such, for instance, has been the sequence of the concepts of particle, force, gravity, field, electromagnetism, atom, quantization, relativity, and so on.

As an example of the significance of these physical concepts, I mention the concept of the *field* and the concept of the *force,* which are directly relevant to our discussion. Physical fields are generated from natural sources, and from the fields originate the forces of nature which cause the transformations of energy (the transformation – the conversion – of energy between its various forms) as is schematically shown in Fig. 1:

L. G. Christophorou, *Emerging Dynamics: Science, Energy, Society and Values*,
https://doi.org/10.1007/978-3-319-90713-0

Source → Field → Force → Energy Transformations

Fig. 1 The physical fields originate from natural sources; via the forces they produce, they cause the transformation of energy between its various forms

Sources of physical fields are electric charges, mass and energy. The physical fields do not occupy a specific space, but they extend in space, including "empty" space; their intensity decreases as the distance from their source increases. The physical fields are invisible; they become "visible" indirectly, through their interactions with, and their effects on, matter.

The Beginning and the Evolution of the Universe

"Is the universe infinite or finite?" Philosophers, theologians and scientists throughout history occupied themselves with this question. Ancient philosophers such as Aristotle maintained that the universe is infinite in space and time, while monotheistic religions considered that the universe is of finite age, that it did not always exist, but it started abruptly "from nothing" at some time in the past.[2] Modern science has strong indications that the universe began to exist in a cosmic explosion 13.8 billion years ago (see, for example, References[3,4,5,6,7,8,9,10,11,12]); we accept this prevailing scientific view in this writing, although there are other, less accepted, scientific views. For example, some scientists argue that the cosmic explosion cannot be verified or disproved definitively, whereas some other scientists maintain that the theory of cosmic explosion does not prove that there was a beginning of time, as the present expansion of the universe may be one phase of an oscillating or cyclic universe.[13,14] There is also the extreme hypothesis of the "multiverse"according to which our universe could be one of an infinite set (see endnotes 12, 14) [15]

The principal scientific evidence that the universe is not eternal, but that it began to exist in a cosmic explosion, in a "big bang", 13.8 billion years ago is the following: (i) the expansion of the universe, (ii) the existence of cosmic background radiation, (iii) the fact that the universe today is not in thermodynamic equilibrium, and (iv) the relative abundance of the different elements, for instance, hydrogen (*H*) and helium (*He*), in the universe.

(i) Nearly a century ago (in 1929), science discovered that the universe is expanding.[16] Measurements by Edwin Hubble showed that the distant galaxies recede from each other with a speed approximately proportional to the distance between them. Thus, the universe is expanding in all directions and it was denser in the past. Given the expansion of the universe, science arrived at the beginning of the cosmos, at the big bang (BB), starting from today's scientific facts. Based on the physical laws as we know them today, science has arrived, gradually progressing backwards in time, to moments when the universe was denser and hotter, until the moment when the universe was *unimaginably small*, *unimaginably dense*, and *unimaginably hot* (according to some theories the temperature of the universe in its first 10^{-43} s exceeded 10^{32} degrees Kelvin). This moment, 13.8 billion years ago, marks the beginning of the universe. There is therefore clear scientific evidence that the universe has a beginning.

(ii) A little later, in 1964, science discovered that there is *cosmic background radiation* evenly distributed throughout the universe, which today corresponds to a temperature of 2.7 K. The uniform distribution of cosmic radiation shows that it concerns the entire universe and that it is the radiation left over when the universe was still very hot (~ 3000 K) and very dense and its main constituent was the thermal background radiation. As the universe expanded, the cosmic background radiation corresponded to lower temperatures up to its present value of 2.7 K. The existence of this cosmic background radiation is a clear indication that the universe began to exist at some time in the past. The observations of Arno Penzias and Robert Wilson on the existence of cosmic background radiation uniformly distributed in the universe were announced in 1964. In 1989 NASA launched the "*Cosmic Background Explorer (COBE)*" satellite, which found that the spectrum of the cosmic background radiation coincides almost entirely with that of an ideal black body at a temperature of 2.725 ± 0.002 K. This observation is amazingly consistent with the predictions of the theory of big bang.[17]

(iii) Since the universe contains whatever exists, it constitutes a closed thermodynamic system which tends toward thermodynamic equilibrium. If the universe were eternal, it would already have been degraded energetically and it would already have ceased to exist. Since the universe today exists and it is not in thermodynamic equilibrium, it cannot be eternal, but it ought to have begun to exist.

To the conclusion that the universe is of finite age, one is led also by considering that the energy of radioactive atoms (nuclei) decreases over time because radioactive atoms are metastable and they decay (are de-excited) automatically, radiating a portion of their energy. If the universe were eternal, there would be no radioactive atoms on Earth today; they would already have been de-excited and they would already have been converted into stable atoms. Similarly, one might observe that if the universe were eternal, the interior (the core) of the Earth would not be hot today; it would have been cooled down.

(iv) The relative abundance of various atoms: hydrogen, *H* (10,000); helium, *He* (1000); oxygen, *O* (6); carbon, *C* (1); all the rest types of atoms (<1) (see, for instance, References[6,11,12,18]); hydrogen and helium are primitive ("αρχέγονα") elements, they were created mostly in the early phases of the universe and they reveal the characteristics of its evolution.

Modern science therefore considers that the cosmic explosion marks the absolute beginning of the physical universe: the absolute beginning of time, space, energy (matter) and change. Time started when space started and energy was created; and from that moment onward started the unceasing perennial change and evolution of the physical universe. The expansion of the universe and the consequent drop in its temperature and density, especially in its first few minutes of age, determined its material composition under the perpetual influence of the forces of nature and the incessant transformations of energy.

In the absolute beginning of the universe the prevailing conditions were extreme. Although we do not know well the forms of energy in the first moments of the universe, we know that in the beginning all was energy, incomprehensible quantities of energy in the form of pure radiation (light)[19,20] under extremely high temperatures (see endnotes 6, 8–12, 15, 18)

The radiative energy at the beginning of the universe was gradually transformed into other forms of energy, other types of radiation and other types of particles and antiparticles (see endnotes 6, 8–12, 15, 18): at first to q*uarks*; a little later, to *nucleons* (protons and neutrons) and to *leptons* (electrons, neutrinos and light particles); and much later, to *atomic nuclei* (from the fusion of protons and neutrons). In just the first few minutes of the universe's life, all the essential basic ingredients for creating *neutral atoms* of matter emerged from the primordial radiative energy. Although the atoms of hydrogen and helium appeared in the first few minutes of the universe's age, the "*atomic era*" followed much later.

When the universe was ~ 300,000 years old and its temperature ~ 3000 K, (see endnotes 5, 6, 11) the universe began to fill with neutral matter, the electrons and the nuclei that existed began to combine to produce neutral atoms. With the disappearance of the electrons and the nuclei, matter began to become transparent to radiation and the light began, ever since, to fill the universe.

Progressively, the energy composition of the universe began to change dramatically; the density of matter began to overtake the density of radiation and ultimately matter (the condensed form of energy) prevailed in the universe. The simple neutral matter (initially in the form of *H* and *He*) became successively more complex and diversified the microscopic and macroscopic composition of the universe. The ceaseless transformations of energy and the resultant perpetual change led to the macroscopic universe, to its wondrous structures, and, on Earth, to the amazing order and organization of biological organisms and to life itself.

Today, the radiation in the universe is a very small percentage of the matter-energy that exists (mainly as cosmic background radiation) and antimatter no longer exists on the macroscopic scale (our entire galaxy consists of only matter and not antimatter (see endnote 15)). The matter and antimatter that existed in the initial stages of the universe were by-and-large mutually neutralized under conditions which led to the dominance of matter as we see today.

How did matter prevail over antimatter? We do not yet have a complete explanation of this asymmetry between matter and antimatter.[21] Across the universe, 99% of ordinary matter exists in the form of hydrogen (*H*) and helium (*He*) (see endnotes 6, 11) while, according to recent discoveries in astronomy and cosmology, *dark matter* prevails over ordinary matter (see endnotes 12, 15)[22,23] The existence and properties of dark matter are inferred indirectly from the effects of its gravitational field. But what is dark energy? Scientifically we still do not know.[24]

Thus, science has led us to a uniquely singular moment, to the absolute beginning of the creation of the universe. Science, however, is not able to explain what caused this beginning and from where the primordial energy in the beginning of the universe came. *At this absolute boundary, we can only leap beyond certainty into the unknown – and possibly unknowable – unaided by the known laws of physics.*

Energy at the Beginning and from the Beginning of the Universe

At the absolute beginning of the universe all was one: **energy**; unimaginable concentration of visible and invisible light. From this initial (primordial) energy came all subsequent forms of energy and matter that have since existed and presently exist. Every new form of energy derives from some other form (or forms) of energy that existed before. Energy comes from energy. It could, in fact, be said that we live in a physical universe of energy where everything is a manifestation of the different forms of energy. In this universe, the unceasing transformations of energy degrade the universe's energy and increase its entropy and disorder, while concomitantly they lead to order and organization; everywhere and always, unceasingly, they differentiate the constitution of the universe and account for the physical phenomena and the universe's behavior and evolution. Energy is today of very different forms and very differently distributed in the universe than it was in the distant past.

Nonetheless, science is unable to respond adequately to the question "What is energy?" We know of course a great deal about the various forms of energy and their interactions, (see endnote 18) but we are unable to answer the question "What is energy?" as we are unable of answering the related question "Where did the initial energy come from?". Science cannot explain how something – the initial energy – can come from nothing. Even the suggestion that the quantum vacuum can be a source of energy,[25] presupposes the existence of this very quantum vacuum and the laws which govern its behavior; it displaces the question, it does not answer the question. As was correctly stated (see endnote 15), the "vacuum" of the physicist differs from the "vacuum" of the philosopher, because the vacuum of the physicist is not "nothing". The theory assumes the existence of the quantum field and the laws of quantum physics. But how have the quantum field and the laws of quantum physics come to be? The automatic genesis of the universe "from nothing" remains scientifically unexplained.

In the initial energy that was created at the beginning of time from nothing was contained all that were necessary for the evolution of the universe and life. We exist as living organisms; consequently, the universe had all the necessary forms of energy for life, at least on this planet. In fact, some scientists argue (see endnote 4) that the values of the so-called constants of nature[26] – such as the speed of light, the Planck constant, the gravitational constant, the elementary electrical charge – have the values they do, because in this way are fulfilled the necessary conditions for the emergence of life. Our existence, they say, explains why these constants have the values they do. This interpretation is generally known as the Anthropic Principle, which introduces a rather teleological explanation of the constants of nature: The universe has been tuned in such a way as to allow the emergence of conscious beings such us, and this is the reason the constants of nature have the values they do. Many scientists, however, do not accept such interpretation of the values of the constants of nature.

The Philosophical Dimension of Energy

The concept of energy, as we mentioned earlier, began with Aristotle in the fourth century BC[27,28,29]. However, even today, the philosophical dimension of energy deserves serious study. Many basic questions relating to energy, while being scientifically defined, extend beyond science and remain unanswered; questions like "What caused the beginning of the universe?"; "Where did the initial energy come from?"; "What unites all forms of energy?" or, even, "What is energy, really?". A philosophical study of such questions (see, for instance, References[30,31]) will contribute to the understanding of the concept of energy and will complement the knowledge that comes from studies by scientists in the physical sciences (see, for example, References[18,32,33]). Energy is the fundamental common in science, philosophy and, as we shall see, theology[34].

We will, therefore, refer first to Aristotle's philosophical views on energy and subsequently to the perspective of the Eastern (Orthodox) Christian Theology on energy (energies) that began to develop eight centuries after Aristotle in the fourth century AD. In so doing, we wish to stress that when we refer to the philosophical dimension of energy we extend beyond science, into "trans-science"[35,36,37], where the questions can be formulated scientifically, but have no scientific answers because they lie in a different domain beyond science.

Aristotle's Philosophy on Energy and Its Relationship to the Current Scientific View

For many centuries, matter (energy) was considered infinite in space and time. Aristotle considered matter eternal (αἰώνια), imperishable (ἄφθαρτον) and unborn (ἀγέννητον) («ἄφθαρτον καὶ ἀγέννητον ἀνάγκη αὐτὴν εἶναι»), he writes[38]. However, if energy-matter is eternal, it would not have had a beginning and there would not have been an abrupt transition from absolute nothing to the world. If the universe started abruptly at a moment, this moment is the beginning of time and there was nothing before it. Therefore, the abrupt transition from absolute nothing to the world means that the energy at the beginning of the world did not always exist, it did not precede the world; it means that energy is neither eternal nor infinite.

Initial energy → ***UNIVERSE*** (All respective forms of energy → Physical fields / Forces → Energy transformations → Perpetual change → Evolution) →***TODAY'S UNIVERSE***

Fig. 2 The initial energy is the source of the original fields and forces that shaped the early universe. Thereafter, all respective forms of energy through the physical fields and the forces that they produce, are transforming perpetually the energy and lead to perpetual change and the evolution of the universe, to its present form (In Fig. 2, and in those which follow, scientific knowledge is presented in **red** , the hypothetical / theological in **blue** and the philosophical in **green** color.)

Aristotle refers particularly to "prime matter" (πρώτη ὕλη (see endnotes 30, 31)[39]), which he considers as the substratum (ὑποκείμενον) of all things and all change, the substratum of "all energetic beings" («ὅλων τῶν ἐν ἐνεργεία ὄντων»). According to Aristotle, the prime matter contains all form (μορφή, εἶδος) potentially (ἐν δυνάμει). The transition from prime matter to form, from potential to energetic being («ἀπό τὸ ἐν δυνάμει εἰς τὸ ἐν ἐνεργεία ὄν»), is the perpetual motion (ἀέναη κίνησις) that takes place between them (see endnotes 30, 31)[40,41]. If, then, we assume that the prime matter of Aristotle corresponds to the initial, primordial energy at the beginning of the universe, and if we consider that Aristotle's motion originates from the perennial change caused by the forces of nature, we recognize that parts of the Aristotelian philosophy are relevant to the modern scientific view. The initial energy and all its later forms through the physical fields and the forces they generate (Fig. 1) are transforming incessantly the universe, and the transformations of energy cause the perpetual change of the physical world and its evolution, and consequently, the transition from the "potential" to the "energetic" being («ἀπό τὸ ἐν δυνάμει εἰς τὸ ἐν ἐνεργεία ὄν») (see endnote 41). All material reality is possible – potentially – as argued by Aristotle, and comes into existence, into energetic beings (ἐν ἐνεργεία ὄντα), over time.

Let us then accept a beginning of energy, time, space and change, the explanation of which is beyond science and, in contrast to Aristotle, but according to modern science, let us accept that the energy-matter is neither infinite nor eternal. Then, we can consider that the initial energy is the source of the initial fields and forces that shaped the early universe. Thereafter, all respective forms of energy through the physical fields and the forces they produce are transforming perpetually the energy and lead to perpetual change and evolution of the universe, to its present form (Fig. 2).

Let us, however, take Aristotle's philosophy on energy a step further and focus our attention on change. Change began when the universe began. Since then everything is continuously changing, perpetually becoming; we are, but we are constantly changing. This view is not far from Aristotle's philosophy about motion. According to Aristotle, the motion is eternal and requires a cause, something to cause motion (see endnote 27). Aristotle further argues that motion causes change, and for something to change it should already exist; therefore, the matter that experiences the change preexists change (see endnote 40). If everything that moves, argues Aristotle, must have been moved by something else (and not by itself) there must be an end point in the course of this infinite succession; an Unmoved First Mover ('Ακίνητον

Πρῶτον Κινοῦν) from which originates all movement (all change) in the world (see endnotes 27, 28). And because the motion is eternal and necessary, the Unmoved First Mover must also be eternal and necessary; he causes change without being changed and is eternal in his energy («ἐνέργεια δέ ἡ καθ' αὐτήν ἐκείνου ζωή ἀρίστη και ἀΐδιος» (see endnotes 28, 41).

However, it is not the motion that is eternal, but the Unmoved First Mover. Nor is the motion preceding the change, but, conversely, it is the energy transformations causing the perpetual change and thus the perpetual motion. And since each energy transformation requires the existence of energy, the energy should be derived from the Unmoved First Mover. Accepting that the initial energy is neither eternal nor infinite, we conclude that Aristotle's motion is not eternal, but it has a beginning. The initial energy contains potentially any kind of form, any kind of energy-matter, every energy transformation and every change and motion.

If, then, we consider that Aristotle's prime matter (πρώτη ὕλη) corresponds to the initial energy from which came all forms of energy, the physical fields and the corresponding forces that cause the transformations of energy and the perpetual change, Aristotle's motion and his view regarding the transition from the potential to the energetic being can be rationalized.

If we consider further that the source of the initial energy is scientifically unknown and that the interpretation of the abrupt onset of time, space and energy is beyond science, then, we can modify Fig. 2 as follows (Fig. 3):

The Philosophical / Theological Dimension of Energy

Centuries after Aristotle – during the 13th century Thomas Aquinas and in the 18th century Gottfried Leibniz, among others – formulated, based on Aristotle, the so-called cosmological argument for the existence of God: The universe must have some cause for its existence and this cause is God[42,43,44,45]. In view of the theory of the cosmic explosion, this argument of causality was formulated as follows: Everything that begins to exist has a cause for its existence; the universe began to exist; thus, the universe had a cause for its existence (see endnotes 42, 44).

The beginning of the universe marks the beginning of the laws of Nature, because before the beginning there was no Nature; there were no laws of nothing. The physical laws do not apply beyond the point where there is no space, time and

Original source (Scientifically unknown) → Initial energy → UNIVERSE (All respective forms of energy → Physical fields / Forces → Energy transformations → Perpetual change → Evolution) → ***TODAY'S UNIVERSE***

Fig. 3 (Figure 2 modified); from the original source, which is scientifically unknown, came the initial energy which is the source of the initial fields and forces that shaped the early universe. Since then all the respective forms of energy, through the physical fields and the forces that they produce, are transforming perpetually the energy and lead to perpetual change and the evolution of the universe to its present form

energy. Where, then, did these laws of Nature come from? Seeking an answer to this question, many extend beyond science and argue that the laws of Nature may be eternal. Such inference would mean the existence of laws for a universe that does not exist. Christianity distances itself still further away from science on this point and accepts etiologically that such eternal laws already exist in the eternal will of an infinite being and are projected in the universe. Thus, what science cannot answer, Christianity interprets in its own way: Whatever were the eternal laws that determined the emergence of the universe must have existed from the absolute beginning and therefore must have originated from a source outside the physical universe, and this source is the eternal God.

Let us recap some conclusions based on what has been said so far before we continue to explore the theological dimension of energy:

- The universe began at some time in the past. Whatever physical laws described its behavior during the first moments of its creation were fundamental. Over time the universe became more complex. Energy and its transformations led to the physical world as we know it today.
- Energy did not always exist; it is neither infinite nor eternal. It was created at the beginning of the universe *from nothing* (*ex nihilo*).
- Scientifically we do not know what caused the beginning of the universe. No scientific theory can bridge the gap between the absolute nothing and the existence of the universe.
- The initial energy in the beginning of time entailed potentially everything that followed.
- Beyond science, it can be argued that if the universe has a beginning, it must also have a Creator[46]. Everything that begins to exist has a cause for its existence; the universe came into existence; thus, the universe has a cause for its existence. The initial energy therefore came from something outside of space and time.

If, then, we accept that the universe was created by a superior being that is outside of space and time, then not only the energy at the beginning of time – the initial energy – was created *ex nihilo* at the moment of the cosmic explosion from the out-of-space and out-of-time eternal being, but also the laws that existed at the beginning of the universe and those that followed since pre-existed in the eternal being and were displayed in the universe. ***If, then, the world is an act of the Creator, the cosmic explosion is the initial contact of the infinite with the finite.*** The Creator is outside of space, time and change – infinite, timeless (eternal) and unchangeable. The finite universe and everything it contains is within space and time; it is constantly changing and it is perennially evolving[47,48].

According to Christianity, the universe was created by God "in the beginning" (Gen. 1:1), from nothing (*ex nihilo*). The universe had a beginning in time and is therefore of certain age. St. Basil the Great (330–379) considered that "the beginning of time is not yet time, but neither a minimum part of it" «ἡ τοῦ χρόνου ἀρχή οὔπω χρόνος, ἀλλ' οὐδέ μέρος αὐτοῦ τό ἐλάχιστον»[49,50]. Similarly, St. Augustine (354–430) considered that "the world and time had the same beginning," that "the world was not created in time, but simultaneously with time"[51,52].

The word ***energy*** rarely appears in the New Testament (it appears with various meanings eight times in the epistles of St. Paul[53], in contrast to the word "light"[54] which is scattered throughout the texts of the New (and the Old) Testament. From the 4th century AD however, the words ***energy*** and ***energies*** crept into the tradition of Eastern (Orthodox) Christian Theology in a singularly important way (see, for example, References[50,55,56,57,58,59,60,61]), in a doctrinal controversy of nearly eleven centuries, at the basis of which lies the exploration of the nature of the "energies" and the knowledge of the "essence" of God and the determination of the relationship and the difference between them[62]. The theology of the Eastern Church makes a clear distinction between the energies (ἐνέργειες) and the essence (οὐσία) of God (see, for example, References 50, 55–61).

Many Fathers of the Eastern Church, especially of the 4th century AD (such as the Cappadocian Fathers St. Basil the Great (330–379), St. Gregory of Nazianzus (329–389?) and St. Gregory of Nyssa (330–395?)); but also others, for instance, St. John Chrysostom (349–407), St. Augustine of Hippo (354–430), St. Cyril of Alexandria (376–444), St. Maximus the Confessor (580–662), referred to the energies of God as "divine attributes" and as God's "life-creating energy found within the whole of creation" (see endnotes 50, 55, 56, 59-61); the Earth gives fruit, but with the energy of God that is in it since the beginning of creation. Similarly, later, St. Gregory Palamas (1296–1359) considered that the creation is the work of energy[63].

Lossky[64,65] writes that God wholly unknowable in his essence is wholly revealed in his energies and that the invisible by nature becomes visible by his energy. According to St. Basil the Great, "his energies fall down on us, but his essence remains inaccessible" («αἱ γάρ ἐνέργειαι αὐτοῦ πρός ἡμάς καταβαίνουσιν, ἡ δέ οὐσία αὐτοῦ μένει ἀπρόσιτος») and according to St. Gregory of Nyssa, God "is known only by his energy, is knowable only through his energies" ("διά μόνης ἐνέργειας γινώσκεσθαι»)[66].

St. Gregory Palamas introduced the term *uncreated energies* of God ("uncreated light"), which he distinguished from the *created energies* (energies in creation, in the physical world), and he referred to the uncreated energy of God as the Divine Grace, explaining that the divine grace is not the essence, but the energy of God (see endnotes 50, 55, 56, 59). It was the common belief of the Fathers of the 4th century AD writes Florovsky (see endnote 66), that the Divine Grace is not separated from God; there is a fundamental distinction, but no separation, between the essence and the energies of God (see endnote 59). Thus, the Eastern Christian Theology makes a clear distinction between the uncreated energies of God and the essence of God. If one accepts the distinction of the energies of God into uncreated and created, the former exist eternally, while the latter have a beginning, they are not eternal.

One can thus consider Fig. 4 as the summary of the perspective of the Eastern Christian theology: the initial energy in the beginning of Creation is the product of the uncreated energies of God; before the beginning of creation there was no physical world (there was nothing outside of God); the universe, the physical world and all the forms of the created energy originated from the initial energy.

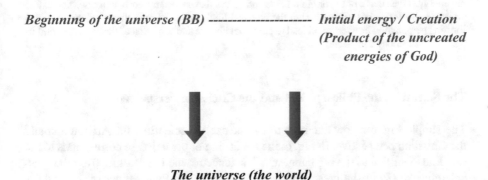

God (essence and uncreated energies)
There was no physical world /
there was nothing outside of God

Beginning of the universe (BB) --------------------- Initial energy / Creation (Product of the uncreated energies of God)

The universe (the world)

Fig. 4 Schematic overview of the perspective of the Eastern Christian Theology

Let us, then, accept a beginning of the universe, which has since been constantly changing and evolving. Let us also accept a Creator, who “in His volition” created the universe *ex nihilo* and that the moment of creation constitutes the initial contact between the infinite and the finite. There was nothing physical before that time when the universe came into existence. The *uncreated energy* of God was beyond space and time, and from it came the initial created (physical) energy at the beginning of time; from the initial energy, over time, originated everything material that existed and presently exists. Everything is therefore energy: *uncreated energy* eternally and *created energy* in the beginning of the universe and since.

It could be even inferred logically that since there were no uncreated energies *in the universe* before the universe came into existence, all energies of God *in the universe* (in creation) have a beginning: the cosmic explosion.

Let us also accept that the most obvious distinction between the Creator and the universe is that the universe is created and has a beginning, whereas the Creator is infinite, uncreated, timeless. The *created (physical) energy* acts in the universe from the beginning of the universe, and, following Eastern Christian Theology, the energies of God are created and uncreated and both are within the human experience[67]. Science does not deal with the uncreated energies or with the essence of God, because they do not belong to its domain. Science deals with – and reveals – the interactions and the effects of actions of the created (physical) energy.

Let us furthermore accept that the cosmic explosion cannot be reversed and that the total energy-matter in the universe cannot gather back to the initial starting point at the beginning of the universe. We can then infer that what was created was created only once and from that came what exists.

Based on the above, the previous schemes can be completed as follows (Fig. 5):

Original source (God)* → *Initial energy and the physical laws* → *UNIVERSE
(All respective forms of energy → Physical fields / Forces → Energy transformations → Perpetual change → Evolution) → ***TODAY'S UNIVERSE***

Fig. 5 From the original source (God) came the initial energy (and the physical laws). The original energy is the source of the original fields and forces that shaped the early universe. Thereafter, all respective forms of energy through the physical fields and the forces they produce are transforming perpetually the energy and lead to perpetual change and evolution of the universe, until its present form (today's universe)

The Scientific, the Philosophical and the Christian Perspective

The simple Fig. 6 below attempts to synthesize the scientific, the Aristotelian and the Christian perspective. In Fig. 6, the beginning of the universe corresponds to the **red** line identified as the big bang, which science reached by starting from the present universe (from the present world). The Unmoved First Mover of Aristotle is indicated by the **green** color beyond the arrow pointing upwards beyond the big bang, which Aristotle reached by starting, like science, from the present world. The Creator God of Christianity in the **blue** color is positioned beyond the big bang and the arrows are pointing downward, from the Eternal God to the big bang.

The Unmoved First Mover of Aristotle is outside the physical world and does not interfere with it. The Creator God of Christianity is eternally beyond time and, while He transcends the world, is always everywhere present in it.

Towards the Whole: Beyond Science

It has correctly been stated that the desire for knowledge is innate. It is, however, necessary to know how well we know what we know, and the kind of knowledge we refer to. The limits of knowledge differ as do the content and the value of knowledge, and the conditions under which knowledge is obtained. Clearly, the data of science differ from those of history and faith; nonetheless, both have undeniable essence and validity. Whether we move within the bounds of science or go beyond science into the domain of philosophy and faith, it is imperative to know how we know what we know, because the quality and value of our knowledge depend directly on the method we used to know: The accuracy of the scientific data, the capability of the experimental method, the validity of the theory or the model, the assumptions, the concepts, and the mathematical, logical, inductive, deductive or any other kind of method we relied upon for knowing.

There are fundamental, self-evident and timeless truths we know well, which are beyond the field of science and oftentimes perhaps beyond logic, truths that do not

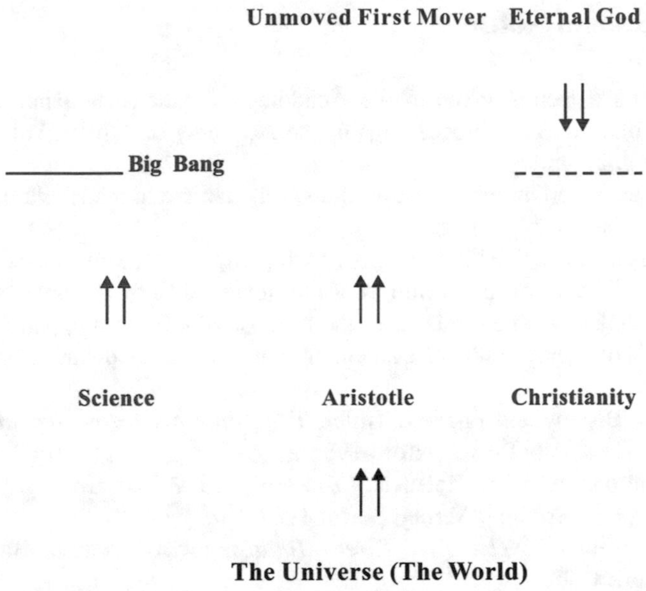

Fig. 6 The Scientific (Big Bang), the Aristotelian (Unmoved First Mover) and the Christian (Eternal God) perspective

need proof. Aspects of human knowledge and existence that determine the quality of life; and I speak of human values, dignity, friendship, reciprocity, love and, yes, about faith in God. This beyond-science-knowledge obviously does not meet the strict scientific criteria, it is not science; however, in my view, it complements our knowledge provided by science.

In summary, we can accept a beginning of the universe and a Creator of the universe, who created the universe *ex nihilo*. From the uncreated energy of God came the initial (physical) energy in the beginning of time, from which emerged everything material that exists. Everything would therefore be energy: *uncreated energy that eternally exists and created (physical) energy that exists from the beginning of the universe.*

I would, finally, conclude that:

1. Science and Christianity – following different paths – reached the common conclusion that there was a beginning of the world; in the beginning, according to science, everything was radiant energy
2. Science, philosophy and Christianity – again following different but complementary paths – reached the common conclusion that, in the beginning of the universe and since, energy is the crucial element of the physical world.

References and Notes

1. Based on a lecture given at the Academy of Athens on January 28, 2014 and published in the Proceedings of the Academy of Athens, Vol. 89, Athens 2014, pp. 27-48.
2. However, in certain religious traditions, for example in the Hindu tradition, the universe had no beginning.
3. There is an extensive bibliography in what concerns the cosmic explosion and the initial stages of the evolution of the universe. Here, are given some references (4-12) which are related to the purpose of this lecture, particularly with the form of energy and matter and their evolution during the initial stages of the universe.
4. John D. Barrow and Frank J. Tipler, *The Anthropic Cosmological Principle*, Oxford University Press, Oxford 1986.
5. G. Contopoulos and D. Kotsakis, *Cosmology – The Structure and Evolution of the Universe*, Springer-Verlag, Berlin 1987.
6. Steven Weinberg, *The First Three Minutes*, revised edition, Basic Books, New York 1993.
7. Stephen Hawking, *A Brief History of Time*, Bantam, New York 1988.
8. John D. Barrow, *The Origin of the Universe*, Basic Books, New York 1994.
9. Paul Davies, *The Last Three Minutes*, Basic Books, New York 1994.
10. Brian Greene, *The Elegant Universe*, Vintage Books, New York 1999.
11. Simon Singh, *Big Bang – The Origin of the Universe*, Harper Perennial, New York 2004.
12. Paul Davies, *Cosmic Jackpot – Why our Universe is just Right for Life*, Houghton Mifflin Company, New York 2007.
13. See, for example, Ian G. Barbour, in *Physics, Philosophy and Theology*, Robert John Russell, William R. Stoeger, S. J. and George V. Coyne, S. J. (Eds.), Third edition, Vatican Observatory-Vatican City State 1997, pp. 22-48. See, also, William Lane Craig and Quentin Smith, *Theism, Atheism and Big Bang Cosmology*, Clarendon Press, Oxford 1993.
14. Paul J. Steinhardt and Neil Turok, *A Cyclic Model of the Universe*, Science **296**, 24 May 2002, p. 1436.
15. Martin Rees, *Before the Beginning – Our Universe and Others*, Basic Books, New York 1998.
16. See, for instance, Reference 11 and John D. Barrow, in *Physics and our View of the World*, Jan Hilgevoord (Ed.), Cambridge University Press, New York 1994, pp. 38-60.
17. http://lambda.gsfc.nasa.gov/product/cobe/
18. Loucas G. Christophorou, *Energy and Civilization*, Academy of Athens, Athens 2011.
19. The term "radiation" is often used to include electromagnetic waves of all wavelengths (of all energies). The term "radiation" is used also to include other types of particles besides photons.

20. Recently, ultra-high energy (>10^{19} eV) cosmic γ-rays were detected, which might have originated from the early universe (Physics Today, May 2010, pp. 15-18).
21. See, for example, References 6 and 15.
22. http://en.wikipedia.org/wiki/Dark_matter
23. There is scientific evidence that ordinary matter constitutes only a few per cent of the mass of the universe and that the majority (96%) of the mass of the universe consists of *dark matter* (23%) and *dark energy* (73%). What is dark energy? Presently we do not know.
24. Some scientists – see, for example, K. Zioutas, D. H. H. Hoffmann, K. Dennell, and T. Papaevangelou, Science **306**, 26 November 2004, p. 1485 – argue that the explanation of dark matter must be sought in "exotic" elementary particles.
25. See, for example, References 4, 12, 15, and 18.
26. John D. Barrow, *The Constants of Nature*, Vintage Books, New York 2004.
27. Αριστοτέλης, *Άπαντα Αριστοτέλους, Φυσικά,* Έκδόσεις «'Ωφελίμου Βιβλίου», Τόμος 5, 'Αθήναι, 1980, Κεφάλαιο Θ', σελίδες 186-226.
28. Αριστοτέλης, Άπαντα Αριστοτέλους, *Μετά τα Φυσικά*, in http://users.uoa.gr/~nektar/history/tributes/ancient_authors/Aristoteles/metaphysica.htm
29. Αριστοτέλης, Άπαντα Αριστοτέλους, *Ηθικά Νικομάχεια*, Εκδόσεις «Ωφελίμου Βιβλίου», Αθήναι, 1979, Τόμος 3, σελ. 425-442.
30. Δήμητρα Σφενδόνη-Μέντζου, «Η Αριστοτελική πρώτη ύλη μέσα από το πρίσμα της Κβαντικής Φυσικής και Φυσικής Στοιχειωδών Σωματίων», in *Ο Αριστοτέλης σήμερα. Πτυχές της Αριστοτελικής Φυσικής Φιλοσοφίας υπό το πρίσμα της σύγχρονης επιστήμης*, Σελ. 61-107, Εκδόσεις Ζήτη, Θεσσαλονίκη 2010.
31. See sources cited in Ref. 30.
32. Crosbie Smith, *The Science of Energy – A Cultural History of Energy Physics in Victorian Britain,* The University of Chicago Press, Chicago 1998.
33. Jennifer Coopersmith, *Energy, the Subtle Concept*, Oxford University Press, New York 2010.
34. Science also can pose the question "*What existed before the big bang?*", but it fails to answer the question; the question rather falls in the domain of philosophy and religion.
35. Loucas G. Christophorou, *Place of Science in a World of Values and Facts*, Kluwer Academic / Plenum Publishers, New York 2001, p. 270.
36. Λουκάς Γ. Χριστοφόρου, *Βήματα στην Επιστήμη και τη Ζωή*, Σύλλογος προς Διάδοσιν Ωφελίμων Βιβλίων, Αθήναι 2009, σελίδες 199-202.
37. See Chapter 6.
38. Aristotle, *Physics* 192a28-29.
39. Aristotle, Physics,192a32-34.
40. Aristotle, *Metaphysics* 1069b 35.
41. Aristotle, *Metaphysics* 1072a19-1072b30, in http://users.uoa.gr/~nektar/history/tributes/ancient_authors/Aristoteles/metaphysica.htm

42. See for example, *Cosmological argument*, http://en.wikipedia.org/wiki/Cosmological_argument, *Ontological argument*, http://plato.stanford.edu/entries/ontological-arguments/
43. William Lane Craig, *The Existence of God and the Beginning of the Universe*, http://www.leaderu.com/truth/3truth11.html
44. William Lane Craig and Quentin Smith, *Theism, Atheism, and Big Bang Cosmology*, Clarendon Press, Oxford 1993.
45. Others (for instance, David Hume and Bertrand Russell) argued that the universe could itself constitute the necessary being (see Bertrand Russell and Frederick C. Copleston, in http://www.bringyou.to/apologetics/p20.htm) http://en.wikipedia.org/wiki/Cosmological_argument
46. If the big bang is the beginning of the universe, the universe must have a Creator, many scientists confess. Apparently, this is the main reason that several other scientists resisted philosophically to accept a beginning of the universe.
47. Before the beginning there is no temporality; in an eternal world, there is no succession but an eternal now.
48. There seems to be a growing recent interest in the so-called "pre-bang universe".
49. Basil the Great "*Hexaemeron*", Homily α΄, II («*Εξαήμερον*», Ομιλία α΄, II.) See, also, Reference 50, p. 102.
50. Vladimir Lossky, *The Mystical Theology of the Eastern Church*, St. Vladimir's Seminary Press, Crestwood, New York 1976.
51. Paul Davis, *Cosmic Jackpot – Why our Universe is just Right for Life*, Houghton Mifflin Company, New York 2007, p. 69.
52. http://en.wikipedia.org/wiki/St._Augustine_of_hippo
53. In St. Paul's epistles the words energy / energies appear eight times (Eph. 1:19-20, Eph. 3:7, Eph. 4:16, Phil. 3:21, Col. 1:29, Col. 2: 12, 2 Thess. 2:9, 2 Thess. 2:11) under various meanings.
54. Light is of course energy, radiative energy.
55. Vladimir Lossky, *The Vision of God*, St. Vladimir's Seminary Press, Crestwood, New York 1983.
56. Vladimir Lossky, *In the Image and Likeness of God,* St. Vladimir's Seminary Press, Crestwood, New York 1985.
57. George Florovsky, Theology **81**, 4, October-December 2010.
58. Peter Chopelas, *The Uncreated Energies. The Light and Fire of God,* in http://joannicius.sovereign.us/UNCREATED%20ENERGIES.htm
59. *Essence-Energies Distinction (Eastern Orthodox Theology),* in http://en.wikipedia.org/wiki/Essence%E2%80%93Energies_distinction
60. Megas L. Farandos, *The Energies of God* http://www.oodegr.com/english/theos/energeies/energeies1.htm
61. Vladimir Lossky, *On the Essence and Energies of God*, in http://solzemli.wordpress.com/2008/12/10/vladimir-lossky-on-the-essence-and-energies-of-god/
62. The theology of the 4th century about the essence and the energies of God found its completion in the 14th century in the Council of Constantinople.

63. St. Gregory Palamas, Reference 50, p. 73.
64. Reference 55, p. 85.
65. Gregory Palamas, Reference 50, p. 86.
66. Stoyan Tanev, "Ενέργεια vs Σοφία: The contribution of Fr. Georges Florovsky to the rediscovery of the Orthodox teaching on the distinction between the Divine essence and energies", in International Journal of Orthodox Theology 2:1 (2011), pp. 15-71. http://www.holytrinitymission.org/books/english/theology_creation_florovsky_e.htm.
67. It should be noted that the Roman Catholic Church, in contrast to the Eastern Orthodox, accepts identity of the substance and the energies of God (see, for instance, References 50 and 59).

Index

L. G. Christophorou, *Emerging Dynamics: Science, Energy, Society and Values*,
https://doi.org/10.1007/978-3-319-90713-0

Printed in the United States
By Bookmasters